Developments in Agricultural and Managed-Forest Ecology, 2

TREE ECOLOGY AND PRESERVATION

Developments in Agricultural and Managed-Forest Ecology, 2

TREE ECOLOGY AND PRESERVATION

by

A. BERNATZKY

Mitglied der Deutschen Akademie für Städtebau und Landesplanung, 6000 Frankfurt am Main 70 (B.R.D.)

ELSEVIER SCIENTIFIC PUBLISHING COMPANY
Amsterdam — Oxford — New York 1978

ELSEVIER SCIENTIFIC PUBLISHING COMPANY
335 Jan van Galenstraat
P.O. Box 211, 1000 AE Amsterdam, The Netherlands

Distributors for the United States and Canada:

ELSEVIER NORTH-HOLLAND INC.
52, Vanderbilt Avenue
New York, N.Y. 10017

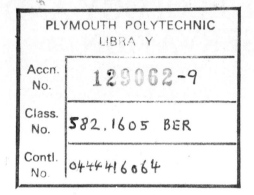

First edition 1978
Second impression 1980

Library of Congress Cataloging in Publication Data

Bernatzky, Aloys.
    Tree ecology and preservation.

    (Developments in agricultural and managed-forest
ecology ; 2)
    Bibliography: p.
    Includes index.
    1. Trees in cities. 2. Trees. 3. Trees, Care of.
4. Arboriculture. I. Title. II. Series.
SB436.B45   635.9'77   78-3623
ISBN 0-444-41606-4

ISBN 0-444-41606-4 (Vol. 2)
ISBN 0-444-41515-7 (Series)

With 227 illustrations and 75 tables

Printed in The Netherlands

CONTENTS

PREFACE

Trees are indispensable to human life. Accordingly we must do all we can to ensure their preservation.

However, all tree preservation depends on knowledge of biological relationships. Only if we give due regard to these, can technical endeavours in this field be successful.

A basic knowledge of the morphology, anatomy, physiology and ecology of plants is assumed, but for the practical worker who is not familiar with every last scientific detail and who may have forgotten much of what he once learned, as well as for the official in a city planning office, the most important data will be briefly reviewed.

This publication deals first with the ecology of trees as discussed in the widespread literature on the subject. Given its limited scope, however, as well as the wide range of research projects conducted the world over, this book cannot claim to treat the subject of tree ecology exhaustively. Many questions of tree ecology still remain to be investigated.

Next, some chapters deal with the effects of trees on man with regard to environmental protection — a subject of great importance.

Lastly, measures for the preservation of trees are discussed which are all derived from extensive experience in this field. Naturally the selection of topics is subjective. The interested reader who seeks further information will find references to basic, as well as to more detailed, literature.

All in all, this publication seeks to stimulate discussion as well as to provide a survey of current research work; it would also hope to serve as an incentive for new investigations in various fields including that of tree surgery. In particular, the wish is to further the practical application of ecological knowledge since without thorough knowledge no practical work can be successful.

The author thanks all those who have helped him in preparing this book, especially the publishers and authors who gave their permission to reprint tables and diagrams and supplied photographs. He feels especially indebted to the two great tree surgery experts in Germany, Michael Maurer and Karl Pessler. The photographs in this book all show their work.

Finally the author would like to express many thanks to Aja Coester (Palmengarten Frankfurt/M) and Bernhard Stummer, Landscape Architect (Neu-Ulm) for their English translation. Without their work, this publication would not have been possible.

1. TREES IN THE HISTORY OF CIVILISATION

Between heaven and earth, between mankind and the stars, lies the king-
dom of trees; between that which gives us food, housing and work on the
one hand (Fig. 1) and that which is incomprehensible and unaccountable on
the other.

Fig. 1. Home for man under trees.

Trees belong to the basic elements of our world. Trees have been present with man from his beginnings. Trees are powerful symbolic figures, even personalities. Many of them reach gigantic dimensions (Fig. 2); a small man is next to nothing compared with them. Others again, as old as the patriarchs of the Bible, become legendary (Plates 1 and 2).

The people of ancient times were far more exposed to nature than we are today. They experienced it directly in all its manifestations and force.

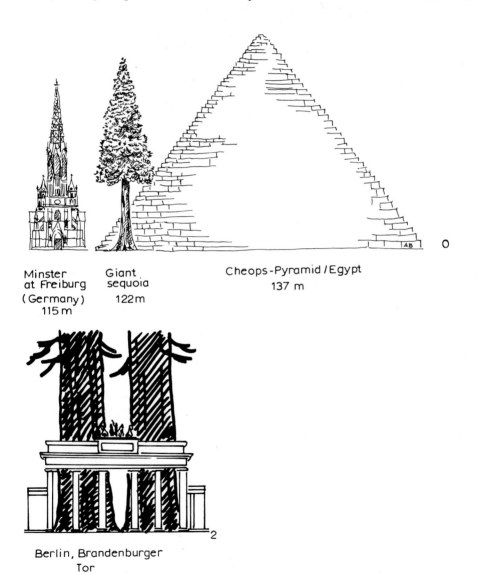

Minster at Freiburg (Germany) 115 m

Giant sequoia 122m

Cheops-Pyramid / Egypt 137 m

Berlin, Brandenburger Tor

Fig. 2. Height and dimension comparison for some buildings with giant sequoia.

4

Plate 1. Cedar of Lebanon, several thousand years old, near Bcharrei/Lebanon (Photo Bernatzky).

Understandably therefore, powerful impulses were felt to emanate from nature and especially from trees. Humans of those early times experienced trees as transcending their world, as supernatural. Their dying in the autumn, their austere immobility in winter and their rebirth in spring made them appear as symbols of powerful gods on whom man felt himself dependent.

Furthermore, trees became the symbol both of man and of the whole universe. The cosmos seemed to early man to be an immense tree looming up out of the ocean and covering heaven and earth with its crown. Traces of this image are to be found in the Sumerian civilisation, with the Prophets of the Old Testament and in the Apocalypse of St. John. The traditional image of the world as the tree *Yggdrasil* in northern mythology is well known. According to the "Edda" the roots of this tree reached deep into the underworld while its green top penetrated the heavens. Thus it connected heaven, earth and the underworld. A link with heaven was possible only by its mediation.

Sacred trees exist in the religions, myths and folklore of all nations (Fig. 3). The tree itself was not worshipped, however, but the God manifesting himself in it. The tree became the dwelling of the deity, as we can tell from reliefs and pictures of Assur, Egypt and Mohenjo-Daro. Buddha (560—480 B.C.) was born under a tree (*Ficus religiosa*) and was also enlightened there. Greek mythology assigned a sacred tree to almost every God. In the rustling of the oak leaves at Dodona the supreme God Zeus manifested himself. At the Capitol in Rome grew the oak of Jupiter. The pomegranate (symbol of matrimonial love and fertility) belonged to Hera, wife of Zeus. Athens' Goddess Athene presented mankind with the olive tree. The oak, the laurel

Plate 2. Linden tree, near Heede/Germany, 700 years old ("12 Apostle-linden") (Photo Maurer).

and the apple tree were attributed to Apollo, God of light. Walnut, willow, myrtle and cedar belonged to Artemis, Apollo's sister. Dionysus was the God of ivy and of the vine. Whenever the Gods wanted to do someone a special favour they turned him or her into a tree. Thus the nymph Daphne was changed into a laurel tree (Fig. 4) and the tormented Cyparissos into a cypress. Philemon and Baucis continued their earthly life as an oak and a linden tree, respectively: this was a reward for the hospitality they had given to Zeus and Hermes. In Germanic and Scandinavian mythology the oak tree belonged to Odin, and the lime tree to Freyja, who guarded matrimony.

Many manifestations of God in the Bible occur in the form of a tree. Trees were planted at altars. In Genesis a tree stands as a mark of decision at the beginning of humanity just as later, on Golgotha, it marks the beginning of a new life. According to tradition, the cross of Christ was made from the tree

Fig. 3. Abraham's Oak near Hebron, Palestine (*Quercus calliprinos*). The oak keeps up the remembrance of the patriarch Abraham. The tree is dying off, unfortunately, as nobody takes care of it. The drawing shows the original state; today only parts of it still exist.

Fig. 4. Metamorphosis of the nymph Daphne into a sweet bay tree (*Laurus*). After a French miniature of the 15th century.

of Paradise (Rech, 1966). Many pre-Christian myths involve the sacrifice or self-sacrifice (by hanging from the branch of a tree) of a mythical personage or divine being: death gives birth to life (Plate 3).

In quite a few traditions, human life is said to have sprouted forth from a tree. Whole clans believe their ancestors to have been trees, therefore regarding the latter as protective Gods. Many fairy tales describe trees growing out of the graves of the dead, symbolizing the care of the deceased for those who remain behind, and even today the trees we plant in cemeteries symbolize the life that does not end in death (Plate 4). Trees became trees of life and trees of destiny. For the tree is "the symbol of life, it represents the living cosmos which forever regenerates itself" (Eliade, 1954).

Plate 3. Sanctuary under trees, Calvaire Quilinen/Britanny (France) (Photo Bernatzky).

Plate 4. *Ficus sykomorus* on a cemetery at Tunamel/Egypt (Photo Werkmeister).

The custom of planting a tree at the birth of a child has survived to the present day. In many German, Slavonic and Lithuanian regions wedding guests carry a green tree ahead of the bridal procession. Many family names derive from tree names. The best known example is that of the three names Linné (*Linnaeus*), Lindelius and Tiliander: three brothers who named themselves after a three-shafted linden tree (*Tilia linden*) at their parental farm. A very old tradition, known throughout Germanic, Slavonic, English, French and Roman countries, is the erecting of the Maypole (Mannhardt, 1963). The point of this ritual is that life is reborn in spring; the pole is supposed to hasten the coming of spring and to further the regeneration and fertility of the fields. In autumn, at the end of the harvest, a green branch or a whole green tree is brought into the barn on top of the last wagon and attached to the roof of the house. On top of the roof beams of a house under construction the builders put a small tree covered with coloured ribbons — symbol of the tree of life which is to protect the home-to-be. The tradition lives on most vividly in the Christmas tree which, like the tree of life, represents the world tree, the source of inexhaustible life. Like many other pre-Christian customs, this one too underwent a Christian transformation. Who remembers today that the ruler's symbol — the sceptre — came from the tree? We are vaguely reminded of this by the ritual in which a staff is broken over a man sentenced to death, symbolizing the destruction of his own life tree. Similarly, the magic wand in fairy tales also originates from the life-giving tree.

This brief review should suffice to show just how great a role trees play in human life and culture all over the world (cf. Rech, 1966; Mannhardt, 1963; Bernatzky, 1973b).

2. A BRIEF HISTORY OF THE EVOLUTION OF TREES

Like all living organisms, trees have an evolutionary process of long duration behind them. It is impossible to say at which exact point in the earth's history trees came into being. Fossils may enlighten us with respect to details relating to the history of the species, but these remnants of the past are few in number. Nevertheless, a rough sketch of the history of plants and trees can be drawn up (Paleobotany) though it will not, of course, provide an answer to each and every question (Fig. 5).

The first plants were very simple organisms (Tallophyta). For many hundreds of millions of years they were the sole representatives of the vegetative world. Some of them, the brown algae living in water, had tree-like structures though they lacked a free trunk.

A milestone in the evolution of plants was their arrival on land. The first land plants, the Psilophytes, came into existence in the Upper Silurian Age about 400 million years before our time. They represented an evolutionary stage characteristic of that period, being not only the oldest but also the simplest vascular plants. In this sense they might be regarded as the tree's oldest ancestors. They consisted of a creeping, forked stock without genuine roots, out of which leafless branches with terminating spore-capsules grew up into the air.

Only 40 million years later — a short time in the history of the earth — in the Devonian Age, plants with roots and leaves arrived on the scene. At the same time, in the Middle Devonian, the Psilophytes vanished. In the Upper Devonian *Lepidosigillaria whitei* and the *cyclostigma* species as well as the genus *Archeopteris* formed stately trees in what is now the North American Continent.

Trees are distinguished from the herbaceous vegetation by the formation of secondary wood. Herbaceous plants possess only "primary wood" consisting of protoxylem and metaxylem. The formation of real (= secondary) wood requires developing tissue (cambium) which continually forms wood on the inside, and bark on the outside, of the trunk.

In the Carboniferous Age which followed (about 330 million years before our time) trees were already forming large forests. In the 70 million years of the Carboniferous, and the 30 million years of the Lower Permian Age almost all of the present-day deposits of bituminous coal developed from the vast forests. In the Lower Carboniferous Age the vegetation of what we now call North America, Spitzbergen, Europe, India and Australia were all essen-

11

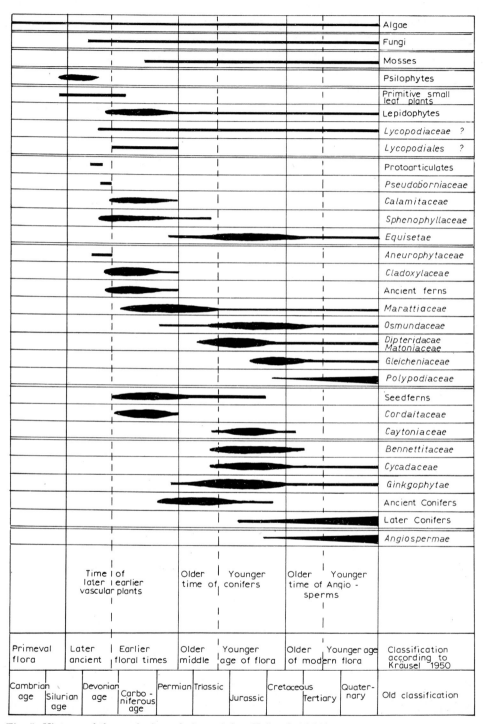

Fig. 5. History of the evolution of plants (after Kräusel, 1950).

tially the same. In this period, woody plants were very common, especially among ferns. However, they differed greatly from today's ferns, the *Lycopodiums* and *Equisetums*, none of which are woody plants but which are herbaceous in structure and without secondary wood.

In contrast, the Paleozoic Age witnessed the start — in varying degrees — of secondary wood growth, so that by that time the *Lycopodiums* and *Equisetums* had become trees. Among the tree-like forms of the *Lycopodium* family were *Lepidodendron* and *Sigillaria*. In the Upper Carbonian Age they included many different species and individuals. *Lepidodendron* formed trees 30 m high with trunks up to 2 m thick. The branchless trunk rose 10—12 m high, then forked into a top with needle-like leaves and long cone-shaped flowers at the end. Some Lepidophyte members of this family had hardly any branches. *Sigillaria* trees were characterized by a wealth of pointed leaves and stem-borne, caulocarpic ones. The name of the mighty *Lepidodendron* was derived from the rhombic scar left by falling leaves. The structure of Lepidophytes would justify their being called "bark trees". In contrast with today's trees, their trunk consisted of only a thin woody part and a very thick bark forming the main part of the trunk: 98% as against 16% in the present-day *Fagus silvatica* and 22% in *Picea abies* (Fig. 6). Even today's herbaceous *Equisetum* developed tree-like forms with real wood in the Carboniferous Period. They were called Calamites, and grew up to 20 m

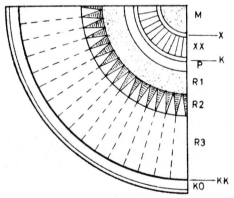

Fig. 6. Simplified schematic cross-section of a sector of a trunk of *Lepidodendron* (after Vogellehner, 1972). M = pith; X = primary wood; XX = secondary wood (real wood); K = cambium; P = primary and secondary phloem; R1 = inner primary bark; R2 = outer primary bark (white: strong-walled mechanical tissue; shaded: thin-walled extension tissue); R3 = secondary bark (white: strengthening tissue; dotted lines: extension tissue); KK = bark cambium; KO = cork tissue. The true wood (XX), that is the secondary vascular conducting and fastening tissue built up by the cambium, takes up only a small part of the whole cross-section, while the bark complex (R1, R2, R3) represents almost all of the trunk tissue.

high. Their wood structure, and especially the structure of their tracheides, could well be compared with those of the later Conifers. The tree-like "genuine" ferns of that time likewise developed secondary wood and hence were trees. A group of them, the Archeopterides, appeared as early as the Upper Devonian Age, showing a distinct trunk structure with tracheides, resembling the later Conifers. Because of their marked differentiation, these trees must be regarded as the first forest-forming trees in the earth's history.

While the trees mentioned so far of the Carboniferous Age had developed their distinct shapes at the end of the Devonian, a group of Gymnosperms called the Cordaites arose during the Middle Carboniferous Age. These grew 10—30 m high and the slender branchless trunks acquired a diameter of 1—2 m. Their wood structure, though not their flowers, anticipated the later Conifers. Their leaves remind one of the Monocotyledons of the Angiosperms. For millions of years the Cordaites were the only seed plants besides Ferns and Lycopodiums; the first genuine Conifers appeared in the Upper Carboniferous Age. This development took place at a time of severe climatic deterioration involving the drying out of large areas of the Northern Continents (end of Carboniferous and Permian Ages). Forerunners of the Conifers in the Upper Carboniferous Age were the *Lebachiaceae*, small trees resembling *Araucaria excelsa*. Next in line were the *Voltziaceae*.

During the Permian and Mesozoic Ages the Conifers were a dominant tree group whereas they contributed only slightly to woody vegetation during Tertiary and Quaternary. The Jurassic and Cretaceous periods witnessed increasing differentiations of the Conifer wood and trunk. The basic structure of present-day Conifer families arose at that time. The latter are the best-known representatives of the Gymnosperms. However, not all Conifers have cones: single-seed fruits occur as well (e.g. *Taxus*). Among the true cone bearers is the *Pinaceae* family (Firs, Spruces, Larches, Pines) whose cones consist of outer scales which tend to drop off and whose axillary pit contains one seed scale with two seeds. In contrast, the cones of the *Araucaria* family bear single scales containing one seed, whereas the *Taxodiaceae* scale often contains more than two seeds; in terms of scale structure, the *Taxodiaceae* occupy a middle position. With the remaining families (the *Cupressaceae*, the *Cephalotaxeae* and the *Podocarpaceae*) the original cone-shape is subject to extensive modification and is often unrecognizable in the adult tree (Kräusel, 1950).

Also similar to the Conifers since the Permian Age is the woody type of the Ginkgos. They played an important role during the Mesozoic Era (Triassic, Jurassic and Lower Cretaceous). Today only *Ginkgo biloba* remains as a "living fossil". The Mesozoic forests were largely made up of Conifers and Ginkgos. The *Cycadeae*, too, are a group of "relics" from the Permian Age. They differ from Conifers in that their smaller and often bulbous or ball-shaped trunks do not grow very tall; also, they grow much more slowly. There are also anatomical differences (little wood, much bark). This group

14

also included the *Bennettiteae*; trees with slender trunks rising several meters high. They are totally extinct today, although at the end of the Mesozoic Era they were widely spread.

Suddenly, during the Middle Carboniferous period, the Angiosperms appeared; most deciduous trees of the present-day belong to this family. Darwin characterized this phenomenon as one of the greatest puzzles in plant history; Kräusel was prompted by it to regard modern plant history as having its beginnings in the Middle Carboniferous Age (Fig. 5), and continuing through the Tertiary until today. The members of the *Ginkgo* family diminish steadily during the Upper Cretaceous; today only *Ginkgo biloba* remains as a "tertiary remnant". Other remnants are *Amentotaxus, Metasequoia, Glyptostrobus, Keteleeria, Pseudolarix* and *Sciadopitys.* Other trees of those ancient times found a refuge in present-day North-America (*Sequoiadendron giganteum, Taxodium distichum* and others).

Plane trees, together with members of the *Moraceae* and *Lauraceae* families as well as poplars and oaks appear to be among the oldest Angiosperms. Some authors assume that Angiosperms first developed in the Northern Arctic. If this is correct they must have spread across the globe in three directions: to North America, to Europe and to Asia. Perhaps in late Eocene two floral regions existed side by side, a permanent one in East Asia which would explain the existence of Tertiary remnants there, and a non-permanent one in the West. There the originally dominant tropical plants slowly moved into the background.

The tertiary plant world evolved into that of the present. However, the habitat of individual forms is subject to change, probably as a result of changes in climate. The changes in the plant world in many northern regions during Tertiary may have to do with a gradual cooling off as a result of continental drift, wandering of the poles or something of this nature. Scientists believe they have a clear picture of the southward migration of plants in North America. All this happened in a time-span of more than 50 million years.

Summary

In the Paleozoic Era trees appear as tree-like forms of *Lycopodium, Sigillaria* and *Lepidodendron*, of *Equisetum* with Calamites and as genuine ferns with *Archeopteris* of the Upper Devonian and Seedferns (Pteridosperms) of Carboniferous origin; also as Cordaites and as the first genuine Conifers.

Dominant in the Mesozoic Era (particularly the Triassic, Jurassic and Lower Cretaceous) are the Gymnosperms (Conifers, Cycads, Bennettites and the *Ginkgo* family). The Tertiary up to the Cainozoic Era is marked by the occurrence of angiospermous floriferous plants, mainly dicotyledonous deciduous trees and (fewer) monocotyledonous ones, as well as relatively few species of Gymnosperms (Conifers, *Ginkgo*, some Cycads).

Present-day taxonomy divides trees into two main groups: (1) Gymnosperms (naked seeds); (2) Angiosperms (covered seeds). There are 850 species of Gymnosperms of which about 750 are Conifers. Most of today's 250,000 Angiosperm species are Dicotyledons as are about 95% of all trees. The Monocotyledons comprise few genuine woody plants. They differ from the Dicotyledons in that their primary conductive bundles, which are distributed all over the cross-section of the stem, are "locked" so that there can be no secondary growth. However, this evolutionary exception did not evolve into a pattern. The tree-shaped Monocotyledons obtained this appearance by other means (the palm trees by primary reinforcement growth, the *Musaceae* by dummy trunks composed of leaf sheaths).

What happened in plant evolution?

Plant history includes processes of decisive significance. Most important was the development of a conducting vascular system and of strengthening tissue (wood) (Figs. 7—9). Wood formation may be compared with steel and concrete, with cellulose resembling the reinforcement and lignin the filling. Wood — i.e. the "secondary wood" — thus becomes the distinguishing characteristic of trees, and because of their wooden trunks, trees have been victorious over other plants in the struggle for light. The factor at work here is growth promotion of the uppermost buds, resulting in the monocorm (single-sprout) tree. The main axis of such a tree grows steadily upwards during the

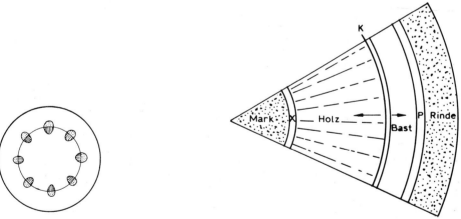

Fig. 7. Conducting vessels in a shoort of gymnosperms and dicotyledons before the beginning of the secondary thickness growth (after Vogellehner, 1972). Dotted = phloem; shaded = wooden part (xylem); circle = later position of the cambium.

Fig. 8. Schematic cross-section of the tissue layers of a typical "wood"-trunk (after Vogellehner, 1972). X = primary wood (proto- and metaxylem); K = cambium; P = primary phloem.

16

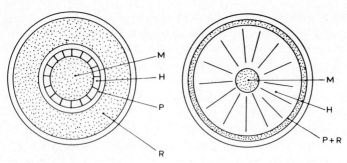

Fig. 9. Schematic cross-sections of a trunk of cycadoid (left) and coniferoid (right) type (after Vogellehner, 1972). M = pith; H = wood; P = phloem; R = bark.

tree's whole lifetime (e.g., Fir). It is interesting to note in passing that trees having a crown of equivalent branches in the mature state always grow monocorm in their youth.

However, the transportation of water and nutrients out of the ground up to the highest parts of the plant requires a pipeline. Ferns developed the tracheides for this purpose. These are elongated dead cells conducting the absorbed water like pipes. Their walls must be braced mechanically, however, since otherwise these dead cells would collapse for lack of turgor. This effect is achieved by the incorporation of lignin (wood) into the membranes. The tracheides thus combine two functions, that of conducting and that of mechanical bracing. Mosses, as yet, have no tracheides but all seed-bearing plants do. The trunks of Gymnosperms (Conifers) are almost entirely made up of tracheides. From the thick-walled tracheid, the path of evolution leads to the long wood fibre with specially thickened walls for stability. From the thin-walled tracheid, on the other hand, the development leads to the trachea which, by dissolution of the cross-walls, forms the vascular system of the angiospermous flowering plants, including deciduous trees and shrubs. Thus the mechanically braced pipeline system traverses the whole tree. In conifers it runs cross-sectionally; in flowering plants it consists of vascular bundles.

The method of reproduction changed essentially also (asexual division — sexual propagation — Gymnosperms — Angiosperms). Reproduction by seed proved to be an especially good adaptation to life on dry land. In the Carboniferous Age the "Seed ferns" had already tried reproduction by seed instead of spores, but the Gymnosperms perfected this method. Their flowers are usually arranged in cones, with males and females separated (exceptions are *Taxus* and *Cupressaceae*). The seeds lie in the cone's axillary pit and are freely accessible (naked seeds). In contrast, Angiosperm seeds are hidden in the ovary which consists of many intergrown pistils. Here the seed does not develop in an open cavity but in the closed ovary which carries the colourful flower in order to attract insects for pollination. After fertilization the petals fall and thick fruit pulp grows around the seed until it is ripe, drops to the ground and germinates. Some 60 million years before our time (Eocene) the development of trees was complete.

3. THE STRUCTURE AND FUNCTIONS OF TREES —
A BRIEF MORPHOLOGY AND PHYSIOLOGY

A tree is a perennial tall-growing woody plant at least five meters high and self-supporting, with a solid trunk and a leaf crown (Fig. 10). The invisible roots are omitted from the definition although they, too, constitute an essential component of a tree. Trunk, crown and roots form an indivisible whole. If one of these parts is in distress, the whole tree is affected (compare Troll, 1973).

3.1. The trunk

That which we call "wood" is a substance of complex structure. It is composed of several different kinds of tissue (Figs. 11—13).

Pith. This is found inside the trunk developing out of the innermost tissues of the vegetational point of the sprout. Its cells build tissue which consists of living or withering parenchyma. The pith traverses, via medullary rays, the entire trunk out to the bark. In a beech tree, for instance, it has a life span of about 40 years.

Wood fibres. These are elongated cells found along the length of the trunk. Mostly inactive, they serve to strengthen the tree (strengthening tissue, sclerenchyma), and represent about 50—66% of the wood substance.

Vessels. In the outer layer of the xylem (wood), at its periphery, are the vessels (tracheae and tracheides). These are the tubular strands of elongated, dead cells embedded in the wood fibres, which convey the water, with nutrient salts dissolved in it, from the soil upwards into the tree crown. This transport of water takes place in the youngest (outermost) layers of wood, often only in the last annual ring.

Xylem parenchyma. In contrast to the dead cells just mentioned, there is the living tissue of the xylem parenchyma. It forms part of the basic tissue in which the vital plant functions (assimilation, respiration, preparation and transport of nutrients, etc.) take place. With its living protoplasm it is able to store organic nutrients which can be released to the tree's growth cells as soon as they are needed.

Medullary rays. These pass through the xylem in the form of small tapes of cells, in a radial direction from the pith to the bark. Together with the surrounding wood parenchyma, they form a cohesive system of living cells which takes care of ventilation, gas exchange and storage. Medullary rays either conduct the downward-flowing assimilates in the sieve tubes of the

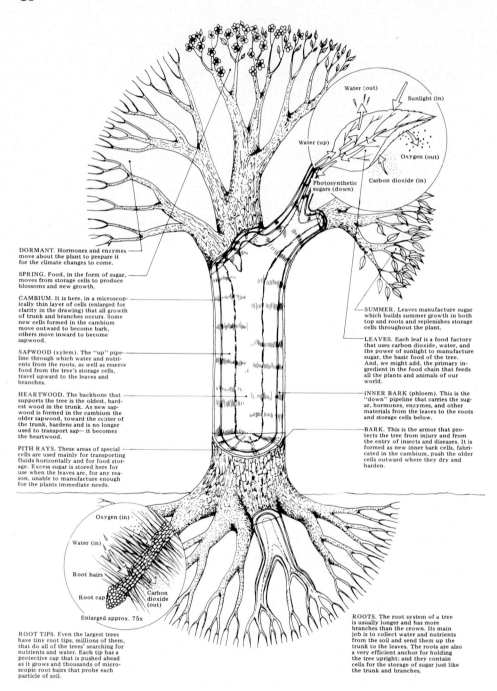

Water (out)

Sunlight (in)

Water (up)

Oxygen (out)

Carbon dioxide (in)

Photosynthetic sugars (down)

DORMANT. Hormones and enzymes move about the plant to prepare it for the climate changes to come.

SPRING. Food, in the form of sugar, moves from storage cells to produce blossoms and new growth.

CAMBIUM. It is here, in a microscopically thin layer of cells (enlarged for clarity in the drawing) that all growth of trunk and branches occurs. Some new cells formed in the cambium move outward to become bark, others move inward to become sapwood.

SAPWOOD (xylem). The "up" pipeline through which water and nutrients from the roots, as well as reserve food from the tree's storage cells, travel upward to the leaves and branches.

HEARTWOOD. The backbone that supports the tree is the oldest, hardest wood in the trunk. As new sapwood is formed in the cambium the older sapwood, toward the center of the trunk, hardens and is no longer used to transport sap— it becomes the heartwood.

PITH RAYS. These areas of special cells are used mainly for transporting fluids horizontally and for food storage. Excess sugar is stored here for use when the leaves are, for any reason, unable to manufacture enough for the plants immediate needs.

SUMMER. Leaves manufacture sugar which builds summer growth in both top and roots and replenishes storage cells throughout the plant.

LEAVES. Each leaf is a food factory that uses carbon dioxide, water, and the power of sunlight to manufacture sugar, the basic food of the tree. And, we might add, the primary ingredient in the food chain that feeds all the plants and animals of our world.

INNER BARK (phloem). This is the "down" pipeline that carries the sugar, hormones, enzymes, and other materials from the leaves to the roots and storage cells below.

BARK. This is the armor that protects the tree from injury and from the entry of insects and diseases. It is formed as new inner bark cells, fabricated in the cambium, push the older cells outward where they dry and harden.

Oxygen (in)

Water (in)

Root hairs

Root cap

Carbon dioxide (out)

Enlarged approx. 75x

ROOT TIPS. Even the largest trees have tiny root tips, millions of them, that do all of the trees' searching for nutrients and water. Each tip has a protective cap that is pushed ahead as it grows and thousands of microscopic root hairs that probe each particle of soil.

ROOTS. The root system of a tree is usually longer and has more branches than the crown. Its main job is to collect water and nutrients from the soil and send them up the trunk to the leaves. The roots are also a very efficient anchor for holding the tree upright; and they contain cells for the storage of sugar just like the trunk and branches.

Fig. 10. Structure and functions of a tree (Chevron Chemical Company).

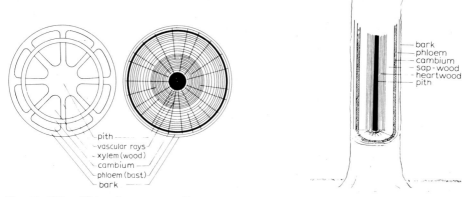

Fig. 11. "Wood"-trunk — cross-section.

Fig. 12. "Wood"-trunk — longitudinal section.

bast radially towards the wood of the trunk and the roots, or store them like the wood parenchyma. Because they sometimes adjoin vascular strands they are able to take water out of these and conduct it from the wooden part outwards. Generally only the outermost wood tissues of the annual rings (sapwood) have active cells which can store reserve substances. The transport of water also takes place in these youngest layers of wood. The older inner wooden parts (the heartwood) provide structural support to the trunk; here the vessels are blocked. Heartwood is mostly darker than sapwood, also more solid, more compact and heavier. Embedded substances (e.g. tannin, phlobaphen) protect it against decomposition. Some trees (hornbeam, beech, birch) have heartwood which does not differ in colour from sapwood. Others again have no heartwood at all (linden, poplar, willow); these trees therefore become hollow as they age.

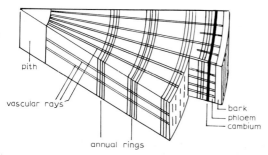

Fig. 13. Segment of "wood"-trunk showing annual rings.

Cambium. The xylem part of the trunk is limited outwardly by the cambium. Consisting of only one layer of cells, this is a particularly important zone of the trunk. It is the zone of growth which lies only a few centimeters or even millimeters beneath the outer surface of the trunk. It consists of thin-walled cells rich in plasm. By cell division it produces new tissues inwards (xylem) and outwards (phloem), but more of the former than of the latter. In the region of the medullary rays it produces parenchyma (basic tissue), so that the medullary rays can pass through from the pith to the bark. The inwardly produced tissue is the secondary permanent wood. The outwardly produced tissue is bast, constituting — together with the layers of cells produced by the activity of a further cambium directly beneath the surface — what is commonly called the bark. The cambium runs parallel to the surface of the trunk as a hollow cylinder, thus forming a ring in cross-section. In temperate zones the growth of the cambium layer is subject to seasonal deviations, whereas in tropical zones it is not. Annual rings thus become visible in the wood structure. They are formed by adaptation to the greater need for water in spring when large vessels are mainly produced, and the different conditions in summer when smaller vessels and strengthening tissue are formed.

Bast. Like the wood, the bast consists of several types of cells differing in form and function. The phloem fibres correspond to the wood fibres; from them the phloem derives its tensile strength, although the phloem fibres are not lignified. The sieve tubes in the phloem correspond to the vessels in the xylem; the former are living cells with unlignified walls. They conduct the assimilates formed in the leaves to the places where they are to be used. In so doing, they also provide the growing roots with the structural and respiratory material they require. In cell structure and function the bast parenchyma corresponds to the xylem parenchyma. The medullary rays also continue right into the bast. Some trees have further particular cells here: excreting cells with oxalic lime in hazel and walnut, milky tubes (gum and caoutchouc), and secreting cells in the resin tubes. Conducting and storage tissue of the bast (soft bast) alternates with the bast fibres (hard bast) in layers. During one vegetation period several layers of hard and soft bast may be formed. In this respect the structure of the bast deviates from the annual rings of the wood. In contrast with the wood which grows outwards unchanged, the outer layers of the bast in time grow too narrow, get torn apart, die and detach themselves. Only the inner bast layers remain uninjured. At the same time, a second cambium forms directly beneath the surface of the trunk, releasing cells by division mainly to the outside. These cells build the bark of the trunk, developing sometimes in the shape of scaled bark (plane-tree) or ringed bark (periderm of the cherry), and so on. The dead, inactive bark constitutes a very important protection against evaporation, change of temperature and damage to the trunk from outside (insects, animals, man).

Damage to the trunk

The frequent instances of damage to tree trunks are generally regarded as harmless. An injury to "a little bit of bark" is accounted of small significance. No attention is paid to the fact that this thin layer at the outer edge of the trunk is one of the most important areas of the tree with regard to its vital processes. Due to interruption of the conducting channels in the cambium and in the xylem close to the cambium, the tree crown and roots can no longer be properly supplied so that their functions deteriorate. The transport and storage of assimilates are disturbed and the strength of the trunk is reduced. The tree has to devote its energies primarily to healing the wound. A wound about 30×50 cm in size takes 10—15 years to achieve complete cicatrization. During this time the tree is more susceptible than usual to infections (fungi, bacteria). Also, the interference in the transport of water and assimilates persists until healing is completed.

All this considerably diminishes the functions and growth of the tree, thus also impairing the tree's capacity for cooling, purifying and renewing the air. In addition, a tree specialist should see to the regular disinfection, wound treatment, watering, fertilization, etc. of the tree during these 10—15 years.

3.2. The tree crown

In the region of the tree crown, the trunk divides into branches, all of which follow the same structural principles as the trunk, having similar vessels, sieve tubes, etc. The trunk serves as static support for the transportation of water and nutrients upwards and assimilates downwards. One of its functions is to position the leaves of the tree crown towards the sun in such a way that they make optimum use of the solar energy, which is in fact the energy required for their vital processes. The leaves are the true productive workshops of the whole tree.

The leaves

The leaves are outgrowths of the shoot axis, i.e. of the branches and twigs. They differ fundamentally from the latter in that their growth is limited in comparison. The shape of the leaf surface is only partly related to its function.

The total surface area of the leaves of a tree is very large. Isolated trees with low branches, such as those in gardens and parks where there is more growing space than in the forest, produce many more leaves than do forest trees. In the case of one 100-year-old beech tree standing free, which was 25 m tall with a crown diameter of 15 m, it was established that the ground area covered by its crown amounted to 160 m^2; its external leaf surface area, however, was 1600 m^2 (Bernatzky, 1967, 1969b), while its "inner" leaf surface area (sum of the intercellular walls of the leaf interior) came to 160,000 m^2 = 16 hectares (Walter, 1950). Only from these figures does the extent of their functions become evident.

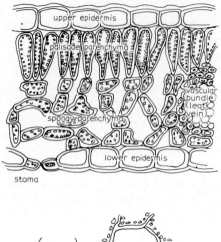

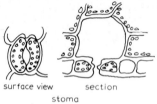

surface view   section

stoma

Fig. 14. Section through a leaf, with detail of stoma.

Every single leaf is composed of three types of tissue that differ in structure and function (Fig. 14): the upper skin (epidermis), the filling tissue (mesophyll) and the vascular bundles. The epidermis generally covers the leaf on the upper side without a gap, whereas the underside is perforated with split openings (stomata). The latter make possible the important exchange of air between the atmosphere and the interior of the leaf. The stomata are surrounded by closing cells whose functions are regulated by light, temperature and moisture. The epidermis cells contain no chlorophyll. Their outer walls are thickened and covered with a delicate, thin but solid skin (the cuticula). Water and gases penetrate it only with difficulty; its functions are protective. Generally the epidermis serves to prevent a rapid drying out of the leaf interior. Between the two leaf surfaces lies the mesophyll, a green filling tissue. It is divided into the palisade parenchyma and the spongy parenchyma. The former is composed of one or more layers of cylindrical elongated cells, the latter of irregularly shaped cells with wide intercellular spaces connected with the stomata. Both types of parenchyma cells contain chlorophyll grains or chloroplasts, the palisade cells containing more than the spongy parenchyma cells. Chlorophyll enables photosynthesis to take place (assimilation). The surface of all the chlorophyll grains in one tree (e.g. a big pear tree) amounts to 2 hectares. The mesophyll also contains certain excretions harmful to the tree, such as surplus lime or calcium oxalates. The leaf veins (nerves) are the continuation of the vascular bundles of the trunk and branches. Towards the upper side of the leaf lie the vessels conveying water and nutrients (xylem) and towards the underside the sieve tubes (phloem) which carry off the assimilates. At the same time, thick-walled fibre cells connected to the veins provide support to the leaf blade. This structure appears in almost all leaves. Deviations are caused by different expositions of the leaves to the sun. Shade leaves are bigger and thinner (they have only one layer of palisade cells) as well as softer.

Function of leaves

Thorough permeation of a whole plant body by water is essential for the normal functioning of its vital processes. As a rule water is absorbed via the roots; by means of transpiration it reaches the furthest tips of the leaves in the tree crown. The sun causes evaporation at the leaf surface. This upward flow of water from the roots is kept in motion chiefly by solar energy. Trees evaporate vast quantities of water (page 55).

A more essential function of the leaf is the production of organic material in photosynthesis (assimilation). This is the most important biochemical process on earth, both qualitatively and quantitatively, since all organic substances as well as the classical sources of energy originate from the CO_2 assimilation of green plants. At the same time, photosynthesis maintains the whole process of "life", which requires a constant supply of energy. In sunlight, one square meter of leaf surface area produces 1 gram of sugar per hour. The annual assimilation yield of the earth's vegetation amounts to more than 100 billion tons of carbon (= more than 100 times the world's coal requirement). The common formula for assimilation reads as follows:

$$6 \text{ mol } CO_2 + 12 \text{ mol } H_2O + 675 \text{ calories}$$
$$\phantom{6 \text{ mol }} 264 \text{ g} 216 \text{ g}$$

$$\rightarrow 1 \text{ mol } C_6H_{12}O_6 + 6 \text{ mol } O_2 + 6 \text{ mol } H_2O$$
$$\phantom{\rightarrow 1 \text{ mol }} 180 \text{ g} \phantom{C_6H_{12}O_6} 192 \text{ g} \phantom{+ 6 \text{ mol }} 108 \text{ g}$$

The steps taken in this process are extraordinarily complicated. Roughly, what happens is that in the process of assimilation water is split photolytically by a reaction of light; this releases oxygen which comes not from the carbon dioxide, but from the water. In the subsequent dark reaction, CO_2 is transformed into a carbohydrate combination with a pentose. The resulting sugar is the basis for the production of other organic materials: cellulose, glucosides, alkaloides, saponines, gum pigments, odorous substances, etc. From the simple sugars fats are also produced. Furthermore, with the help of nitrogen from the soil, fatty acids form albuminoids which are a main constituent of protoplasm. Sulphur and phosphorus also take part in building up albuminoids. Accordingly, the performance of the above-mentioned beech tree (page 21) per hour of sunshine is as follows:

CO_2 intake	2352 g (total CO_2 from 4800 m³ of air)
H_2O intake	960 g
$C_6H_{12}O_6$ production	1600 g
O_2 output	1712 g

Converting these figures for a green area with trees, shrubs and turf, this means that 1 hectare, whose total leaf surface is about 5 hectares, in its fully-grown state draws 900 kg CO_2 from the air in 12 hours, and at the same time releases 600 kg O_2 (Bernatzky, 1967; from the figures according to Walter, 1950; Strasburger, 1971). A man's oxygen needs are met from the O_2 production of 150 m² of leaf surface per year (Walter, 1950). Therefore this oxygen, so essential for life, is, as it were, a by-product of photosynthesis.

The dubious practice of pruning trees

The great importance of trees and their leaves is obvious from the section above. However, the slightest damage to their foliage reduces their rate of metabolism and weakens the growth of all parts of the trees, including the roots, which need the assimilates from the leaf crown. Therefore it is essential for a tree to develop and maintain a full, healthy leaf crown, so that it

Plate 5. Such trees are worthless (Photo Maurer).

will thrive and function well in the service of man, and it is incomprehensible that again and again trees are being relentlessly cut back. They can of course survive these violent attacks but only under great strain, with risk of disease and loss of function (Plate 5).

3.3. The roots — the "underground" tree

While the trunk and crown of trees are visible and taken care of reasonably well, the root area is generally overlooked and neglected. However, the root system is the underground equivalent of the tree crown. Anchoring the tree in the ground and intake of nutrients and water are carried out by the roots, but the actual absorption of water and mineral salts goes on only at the root tips, in the zone of the root hairs (Fig. 15). The tip of the fibrous root is formed by the root cap which covers the true root tip with its vegetative cone, just as a thimble covers the finger. The oldest cells steadily die, produce mucilage and so make it easier for the root tips behind them to grow on into the ground. Close behind the vegetative cone is the zone of elongational growth (5—10 mm long), and behind that the zone of root hairs. The latter are protruding epidermis cells covered with mucilage (ca. 0.15—8 mm long). They have a short life. The root hairs enlarge the root surface up to 18 times. This in turn considerably expands the areas of contact with the soil which is

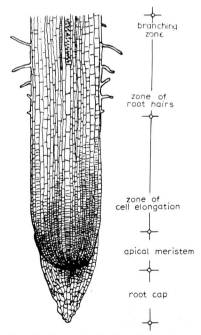

branching
zone

zone of
root hairs

zone of
cell elongation

apical meristem

root cap

Fig. 15. Longitudinal section of a root-tip.

26

important for the intake of water and salts. This is a very complicated process. By diffusion, or osmosis, water enters the surface cells of roots, particularly the root hairs. This is possible only when the cell sap has a higher concentration than the soil water. High osmotic pressures are required to overcome the tendency of water to remain in the soil.

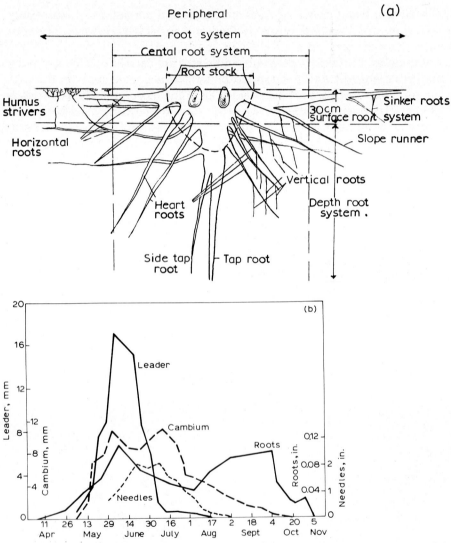

Fig. 16. (a) Root system (after Melzer, 1962). (b) Seasonal variations in amount and duration of shoot growth (elongation of the terminal leader and of needles), cambial growth, and root elongation. From Physiology of Trees by P.J. Kramer and T.T. Kozlowski. Copyright 1960 by McGraw-Hill, New York.

Direction of root growth

After germination all roots grow positively geotropic (Kozlowski, 1971a; Zimmermann and Brown, 1971; Malcolm, 1974). Sometimes these tap roots die and other roots develop parallel to them. These are called heart roots (Fig. 16a). Some tree species have only this type of tap or heart root, others develop lateral roots, but even roots which at first grow positively geotropic may modify the root pattern as a result of soil texture, water availability and overall nutrition. Melzer (1962) calls the disposition of the root system a species characteristic but considers its actual formation dependent upon location.

Root growth is stimulated by water, temperature and nutrients (Table 1). Mechanical obstructions are by-passed by primary as well as lateral roots which, after getting around the obstacle, continue to grow in the original direction. Root growth occurs periodically. "The periodicity of root growth may be summarized as follows: not all roots of a tree grow at any one time; while some roots are growing, others are quiescent. In many angiosperms and gymnosperms of the temperate zone, there is a peak of active root growth in spring which may begin before, during or after shoot growth. This activity is followed by a reduction of growth during summer in both rate and number of roots produced. In the fall, there may be another peak of activity, after which the majority of roots become quiescent with the onset of winter and low soil temperatures (Fig. 16b). Individual roots show cyclic periods of growth indicating some endogenous mechanism of control; however, this control may be modified by changes in the external environment. Seedlings

TABLE 1

Maximum daily increase of single roots (mm) of young plants from different tree species (after Lyr, 1967 in Lyr et al., 1967)

Tree species	Increase (mm)	Author
Robinia pseudacacia (locust)	56	Hoffmann, 1965
Populus euramericana cv. Sacrau (poplar)	50	Hoffmann, 1965
Quercus borealis (red oak)	18	Hoffmann, 1965
Betula pendula (white birch)	15	Hoffmann, 1965
Pseudotsuga taxifolia (douglas fir)	16	Hoffmann, 1965
Larix leptolepis (japanese larch)	10	Hoffmann, 1965
Picea abies (spruce)	8	Hoffmann, 1965
Pinus sylvestris (scotch pine)	12	Hoffmann, 1965
Pinus taeda (resinous wood pine)	3—5	Barney, 1951
Pinus taeda	25	Reed, 1939
Pinus echinata (shortleaf pine)	25	Reed, 1939
Malus hybrid (apple tree)	3	Rogers, 1939
Acer saccharum (sugar maple)	ca.1	Morrow, 1950

are more responsive to environmental change than older trees" (Zimmermann in Zimmermann and Brown, 1971).

The largest number of roots is found in medium-textured soils in the area 90 cm below ground level (Zimmermann in Zimmermann and Brown, 1971), and according to Walter (1960) within the uppermost 50 cm. The fibrous absorbing roots generally grow in the upper 15 cm layer of soil. This is the general case; however, some roots reach far greater depths. At an open location the extension of the lateral roots may be two to three times the radius of the crown, even though the majority of absorbing roots will stay within the area circumscribed by the periphery of the crown. The length of all roots taken together may reach 100 km or more. The mass of the roots represents from one third to one half of the mass of the tree (Bonnemann and Röhrig, 1971).

Reduction of root growth may be caused by

lack of light (shading),

removal of branches, followed by reduction of assimilation and corking up of the root tips; complete defoliation of a tree may cause rapid death of most of the small feeding rootlets (Kozlowski, 1972),

lack of air in the soil (overfilling, flooding),

overdoses of nitrogen,

competition with roots of other plants.

The results of root reduction are diminished absorption of nutrients and water, and increased danger of death through drought and windfall. A lawn particularly retards root growth and the upper parts of the tree do not thrive. Investigations on four-year-old apple trees with different ground surface conditions revealed interesting results (Table 2). The severe retardation,

TABLE 2

Root mass and overground parts of apple trees with various kinds of ground surface treatment in g dry weight (after Oskamp in Walter, 1960)

Depth of soil (cm)	Loosened (hoed) soil (g)	Covered with straw (g)	Beneath lawn (g)	Mulched with grass (g)
0—23	1543	893	129	228
24—45	749	1233	207	327
46—67	261	212	25	67
68—90	62	55	7	20
Weight of the above-ground parts (kg)	11.34	11.52	0.89	2.1
Weight of the under-ground parts (kg)	2.61	2.39	0.37	0.64

TABLE 3

Root system of some trees (after several authors)

___

Generally having a tap root system

Abies alba	*Pinus sylvestris*
Carya illinoensis	*Pyrus communis*
Carya ovata	*Quercus alba*
Fraxinus excelsior	*Quercus macrocarpa*
Juglans nigra	*Quercus petraea*
Juniperus communis	*Quercus robur*
Juniperus virginiana	*Sorbus domestica*
Larix decidua	*Sorbus torminalis*
Larix kaempferi	*Sophora japonica*
Liriodendron tulipifera	*Ulmus glabra*
Maclura pomifera	*Ulmus laevis*
Pinus palustris	*Ulmus minor*
Pinus ponderosa	

Generally having a lateral root system (large, shallow and flat spreading below the surface roots)

Acer campestre	*Larix laricina*
Acer saccharinum	*Liquidambar styraciflua*
Acer saccharum	*Malus silvestris*
Alnus incana	*Nyssa sylvatica*
Betula papyrifera	*Picea abies*
Betula pendula	*Picea omorica*
Betula pubescens	*Pinus banksiana*
Catalpa species	*Pinus strobus*
Elaeagnus angustifolia	*Populus*
Fagus grandifolia	*Salix*
Fagus sylvatica	

Having an intermediate root system (wide spreading and deep lateral roots)

Acer negundo	*Prunus avium*
Acer platanoides	*Pseudotsuga menziesii*
Acer pseudoplatanus	*Quercus borealis*
Aesculus hippocastanum	*Quercus pseudoturneri*
Caragana arborescens	*Robinia pseudacacia*
Carpinus betulus	*Taxus baccata*
Fraxinus pennsylvanica	*Tilia americana*
Ginkgo biloba	*Tilia cordata*
Gleditsia triacanthos	*Tilia euchlora*
Pinus nigra	*Tilia tomentosa*
Platanus hybrida	*Tilia platyphyllos*
Platanus occidentalis	

by the grass roots, of the roots and of the parts above the ground is clearly evident.

It is difficult to classify the rooting characteristics of different species of trees. Only with caution can the classification in Table 3 be used, and it must be repeated that the root characteristics mentioned can always be modified by repeated transplanting, by particular site and soil conditions, by obstructing layers in the soil, etc. *Pinus sylvestris*, for instance, can develop five different types of roots, from a tap-root to a lateral-root system, in only 30 cm of soil (Vomperskij, 1959, in Kozlowski, 1971a). *Eucalyptus*, in dry areas, develop a long tap root with few poorly developed laterals. On favourable sites, however, they develop a shallow fibrous root system (Zimmermann and Grosse, 1958, in Kozlowski, 1971a, further examples are given). It is to be noted that generally every healthy, undisturbed and well-sited tree produces, without our assistance, the root system which suits it, as long as we do not put obstacles in its way. A tree can even compensate for the effects of prevailing winds.

3.4. Aging of trees

Compared with other living beings, trees may grow unusually old. The maximum age limit varies with individual tree species; a fact which indicates genetic determination. However, the existence of external factors (storm, fire, diseases) and physiological ones (unfavourable location) makes it far from easy to determine the influence of internal factors on the life span of trees. Generally it has been observed that speedy tree development is coupled with early aging and a short life span. Lyr et al. (1967) and Meyer (1965) deal extensively with this subject.

Important organs such as the leaves (and needles of conifers) age and die relatively soon. Newly developed wood, too, lives only for a limited period, and tracheae, tracheides and wood fibres live only a few days or weeks. Embryonal tissues such as cambium and vegetative points may continue to grow without interruption. If shoot tops are cut off, the trees grown from them may get older than the mother tree from which the cuttings were taken. Meristems capable of division can repeatedly produce new cells, tissues and organs. They are potentially immortal, so long as no other factors cause their death. However, the tree as a whole does not remain in its juvenile state, but passes through different phases (Fig. 17).

Of course our present knowledge about the process of aging is not adequate to enable us to explain it sufficiently. Paraphrases such as "decreasing vitality" and other definitions are used for it. Human beings regard aging as irreversible but things look different in the plant world. With ivy, for instance, the juvenile form of the leaves was achieved again by grafting an aged shoot with typical mature leaves onto a young plant. On the other hand, cuttings of lateral twigs generally maintain their plagiotropic growth especially if they come from aged plants, particularly from conifers. Cuttings

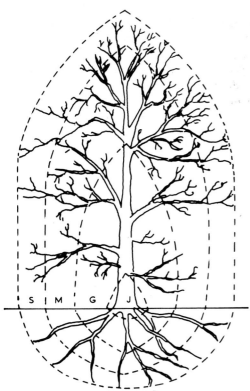

Fig. 17. Illustration of the hypothetical zones of development of a tree. Scheme after Lyr, with coat-like structure showing growth stages (Lyr et al., 1967). J = juvenile stage; G = growing stage; M = mature stage; S = senescent stage.

taken from different heights of the mother plant behave differently. Those originating from the basal area show youthful characteristics, in contrast with those from the top.

The aging of vegetative tissue is greatly affected by correlative influences which also affect meristems and may alter their physiological behaviour. Shoot growth is determined by hormones. Dominating shoots such as top or leading shoots check the growth of subordinate lateral shoots and force them to grow at a slanting angle (plagiotropically). Also, auxines developing in the top shoot cause the formation of inhibitors; the opposing influences of growth substances and inhibitors in top and lateral shoots determine the latter's growth and form and finally fix the shape of the tree crown. In this process, many buds do not sprout at all (sleeping eyes, preventive buds). Only a disturbance of the hormonal economy (dehorning, isolating, pathological injuries) can neutralize the inhibitory effect and cause the buds to sprout. New buds will also develop from callus exuberance (adventive buds). The sprigs rising from them, which possess a juvenile character, will compete with the old parts of the tree. They get a better supply of water and mineral salts. With increasing numbers of growing shoot-ends in the crown the production of inhibitors rises and with it the mutual inhibition of twig vegetation. In addition, the lines for the transportation of water and nutrients get longer as the tree becomes larger. Together with these inhibitions to growth, the vegetative energy of the top shoots is reduced while the lower branches are thriving. All the energy now goes there, while the upper parts of the crown receive less and less so that they finally die away. Storms break them off and fungi enter the tree through the wounds. Its further fate depends on the resistance of the wood and on the presence of antibiotics stored in the wood such as heartwood toxins and tannins, though with increasing age these lose some of their effectiveness by autoxydation.

The extinction of earlier species in the course of the earth's history was due mostly to inadequate adaptation to environmental changes and to the appearance of either natural enemies or competing species. Thus a tree does not die as a result of the loss of vital organs or because the potentially immortal vegetative point grows older. Its death is caused by increasing correlative inhibitory effects in the crown, and by cumulative difficulties connected with the supply of water and nutrients, and the occurrence of pests and diseases (Table 74, page 318).

4. TREES IN THEIR ENVIRONMENT

All living beings depend on their environment and on their interaction with it. No one exists as a solitary being in nature; instead, all organisms interrelate with other organisms. The task of ecology is the registration of these interrelations (and their effects) between various organisms and their environment, that is, of all the organic and inorganic factors resulting from a particular location. Tree ecology focuses on trees as elements of the highly intricate organic structure of the plant world.

The environment of plants includes the atmosphere, the hydrosphere, the lithosphere (including the soil) and the biosphere. The individual organism's relations with and reactions to environmental factors are defined as auto-ecology. This is concerned with vital functions and their regulation by variable external factors. Autoecology therefore employs comparative physiological methods, focusing on the homogeneous organism of the whole plant in nature rather than on its individual functions. Synecology, or the ecology of groups or organisms, deals with the effect of the environment on the whole pattern of vegetation on earth, or in certain areas, and with the competition between individual plants in a group, such as may be found in a park or a forest. Synecology mainly analyses vegetative structures and the spatial alterations of environmental factors.

Environmental influences jointly and continuously affect the plants. Figures 18—20 clearly show the network of influences as well as the plants' reactions. Figures 18 and 20 show how different circumstances may lead to the same result; in this instance to a modest annual wood production. Figure 19 shows a similar network, except for the relationships that are important prior to the growing seasons (Fritts et al., 1971).

Ninety-nine percent of all living beings on earth (biomass) are plants (phytomass). Only the latter are able to receive the energy of solar radiation from space, to store it up and make it available, via the food chain, to human life. This transformation takes place in metabolic cycles, of which there are two basic types: the gaseous cycle with the circulation of carbon, oxygen and water; and the earthbound circulation of mineral bio-elements. Nitrogen circulation occupies an intermediate position. Though plants obtain nitrogen from the soil, the main supply of this element is found in the atmosphere where it exists as a gas.

Certain terms have been coined to describe the various degrees of inter-relation between organisms and their environment. A limited, uniform sector

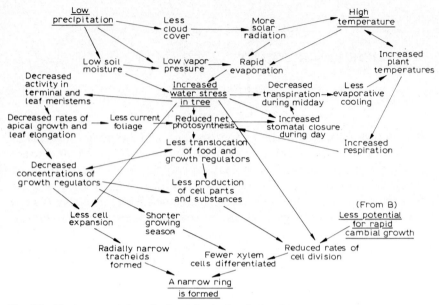

Fig. 18. Environmental and physiological relationships important during the growing season, which will lead to the formation of a narrow ring in conifers growing on semi-arid sites.

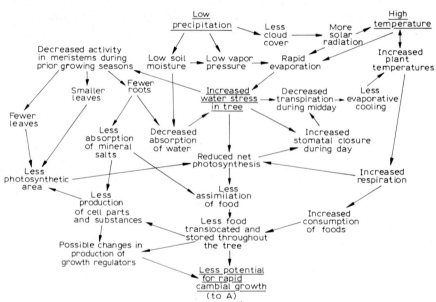

Fig. 19. Environmental and physiological relationships important prior to the growing season, which will lead to the formation of a narrow ring in conifers growing on semi-arid sites.

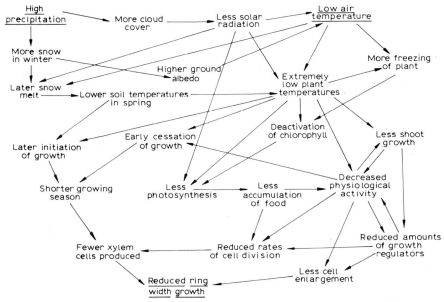

Fig. 20. Relationships that are opposite in effect from those shown in Figs. 18 and 19 which, under certain circumstances, may contribute to narrow rings. (Figs. 18—20 after Fritts et al., 1971.)

of the biosphere is called an ecosystem or biogeocenose. According to Ellenberg (1971) (Fig. 21) an ecosystem is an entirety of an effective system ("Wirkungsgefüge") of living organisms and their inorganic environment, such a system being to a certain degree capable of self-regulation. A forest, a park or a lawn are such ecosystems. While the interrelations always exist jointly they can only be explained separately. Authors have employed different ways of dealing with the matter. This publication has to assume a basic knowledge of plant ecology. A large number of publications on this subject is available (e.g. Odum, 1971, with numerous references; Larcher, 1973a).

4.1. Basic elements

4.1.1. Sunlight as the source of energy

For trees and all autotrophic plants the stream of energy radiated by the sun is the essential source of life. Light is essential for
photosynthesis,
the differentiation of vegetable tissues (photomorphogenesis), e.g. in combination with the water, the production of light- and shade-leaves,
phototropical effects (phototropism), e.g. trunk contortion and inclination as a result of light from one side only,

36

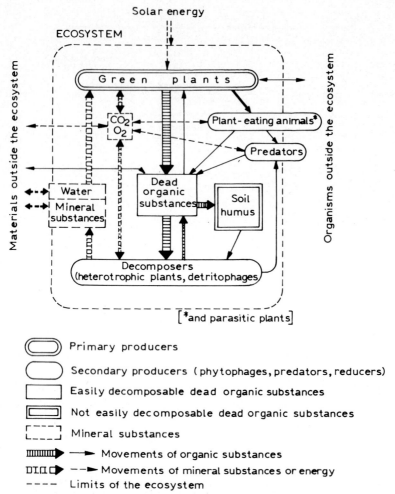

Solar energy

ECOSYSTEM

Materials outside the ecosystem

Organisms outside the ecosystem

Green plants

CO_2
O_2

Plant-eating animals[*]

Predators

Water

Mineral
substances

Dead
organic
substances

Soil
humus

Decomposers
(heterotrophic plants, detritophages)

[*and parasitic plants]

⬭ Primary producers

⬭ Secondary producers (phytophages, predators, reducers)

▭ Easily decomposable dead organic substances

▣ Not easily decomposable dead organic substances

⌐‾¬ Mineral substances

▦▶ ➞ Movements of organic substances

▣▷ --➤ Movements of mineral substances or energy

---- Limits of the ecosystem

Fig. 21. General scheme of an ecosystem. The individual components and processes may
be of varying importance. (After Ellenberg, 1971.)

the effect of different day lengths on a number of periodic processes (photo-
periodism),
photoinductive processes such as the induction of germination, flowering,
and leaf fall, and also for the seasonal change between activity and dormancy,
finally also for carrying out remote sensing.
Of the sun's radiation only about 45% is received by the plants, this amount
varying with latitude and altitude above NN. Apart from this the reflection
of the arriving light on surfaces of different qualities is of some importance.
A reflex-number of 0.3, or 30%, indicates that the soil or surface concerned
reflects 30% and absorbs 70% of the incoming light (Table 4). Inside a tree

TABLE 4

Reflex numbers of different surfaces in the visual spectral range (after Geiger, 1961) in %

Fresh snow cover	80—88
Cloud surface	60—90
Older snow cover	42—70
Fields, meadows	15—30
Heath and sand	10—25
Forests	5—18
Ocean surface	8—10

crown the supply of light received by the leaves depends on their position and on the thickness of the foliage (Fig. 22). Different species of trees have different minimum light requirements (Table 5). The smaller the percentage shown, the greater the tree's shade tolerance (Table 6). This is also affected by other factors, however. It is raised where general growing conditions are favourable. On the other hand, if there is considerable lack of light trees suffer far more from infestations of insects and fungi (Bonnemann and Röhrig, 1971). The diminution of light intensity leads to a decrease of substance production, noticeable at first in root reduction (Table 7). Vertical growth is generally not reduced but sometimes even increased (compare Coombe, 1966). Solar radiation on a forest floor is 2% of the radiation (set at 100%) on the upper surface of tree crowns (Larcher, 1973a). This strong

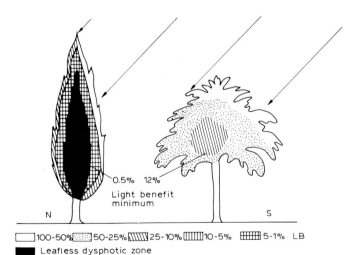

Fig. 22. Relative light benefit in various trees on clear days in July and August at noon (Larcher, 1973a). Light benefit = radiation intensity on various points of the tree crown given in percent of the open-space light intensity.
Left, dense Cypress crown; right, sparse crown of an olive tree.

38

TABLE 5

Light minimum for different species of trees in solitary position and in a close stand (after Wiesner and Hesselmann, in Walter, 1960) in %

Species of tree	Solitary position	Close stand
Buxus sempervirens	0.9	—
Fagus sylvatica	1.2	1.6
Aesculus hippocastanum	1.2	1.7
Carpinus betulus	—	1.8
Acer platanoides	—	1.8
Acer campestre	2.3	—
Populus alba		6.7
Populus nigra		9.1
Pinus sylvestris		10
Betula pendula		11
Populus tremula		11
Sorbus aucuparia		12
Fraxinus excelsior		17
Larix decidua		20

decline is the result of foliage density and leaf patterns. Foliage density is registered by the leaf area index, which indicates how much larger a tree's total leaf surface area is in relation to a particular ground area. Normally these figures compare square metres of leaf surface to square metres of ground surface. The damaging effect of light — especially its UV content — on bacteria and fungi is well known. On unshaded ground this effect extends to micro-organisms of the soil.

Because trees do not react uniformly to alterations in the photoperiod, Nitsch (in Lyr et al., 1967) gave the following classification (Table 8):

Type A (*Populus*) long day results in continuous growth, short day results in dormancy,

Type B (*Quercus*) long day results in periodical growth, short day results in dormancy,

Type C (*Juniperus*) short day brings no dormancy,

for types A, B and C the long day blocks dormancy,

Type D (*Syringa*) long day does not block dormancy.

4.1.2. Heat economy (budget)

Solar radiation is also the source of energy for the heat economy. The uptake of heat energy is counteracted by the output of longwave heat radiation, resulting in a loss of energy. The radiation balance is positive when irradiation exceeds emission (during daylight) and negative, when longwave radiation prevails (mainly at night and in frosty weather). Heat affects photosynthesis,
respiration,

TABLE 6

Shade tolerance of some trees (after Baker, Lyr and other authors)

Very shade tolerant

Abies balsamea	Acer saccharum
Taxus baccata	Carpinus betulus
Thuja plicata	Cornus florida
Tsuga canadensis	Cornus mas
	Corylus avellana
	Fagus sylvatica
	Fagus grandiflora

Shade-tolerant

Abies concolor	Acer pennsylvanicum
Picea glauca	Acer rubrum
Picea rubens	Alnus glutinosa
Picea sitchensis	Fraxinus excelsior
Pinus nigra	Fraxinus ornus
Pseudotsuga taxifolia	Tilia americana
	Tilia parvifolia

Intermediate

Picea abies	Betula allegheniensis
Pinus cembra	Fraxinus americana
Pinus lambertiana	Quercus alba
Pinus monticola	Quercus borealis maxima
Pinus strobus	
Sequoia sempervirens	

Shade-intolerant

Pinus ponderosa	Betula papyrifera
Pinus resinosa	Liriodendron tulipifera
Pinus taeda	

Very shade-intolerant

Larix decidua	Betula pendula
Larix laricina	Betula populifolia
Pinus banksiana	Populus tremuloides
Pinus palustris	Robinia pseudacacia
Pinus silvestris	

warming up of soil, air and plants,
evaporation,
together with light and water it creates the various climates and so helps
determine the distribution of vegetation on earth.

40

TABLE 7

Sprout length, diameter at the root collar, dry weight and leaf surface of two-year-old linden and maple trees at different light intensities (after Röhrig, 1967)

Relative light intensity (%)	Sprout length (cm)	Diameter (mm)	Dry weight (g)				Leaf surface (cm^2) per plant
			Leaves	Stems	Roots	Total	
Tilia cordata							
100	15.0	4.5	0.6	0.5	1.6	2.7	18.6
78	17.2	4.8	0.6	0.7	1.2	2.5	20.6
24	10.2	3.8	0.3	0.3	0.5	1.1	20.1
8	13.0	3.7	0.3	0.4	0.6	1.3	26.6
1	7.4	2.3	0.1	0.1	0.1	0.3	13.6
Acer pseudoplatanus							
100	27.3	5.6	1.8	2.6	3.4	7.8	27.6
78	20.5	4.2	1.1	1.4	1.7	4.2	24.1
24	16.2	4.1	0.5	0.9	1.0	2.4	22.6
8	14.4	3.6	0.4	0.7	0.8	1.9	26.9
1	11.0	1.7	0.1	0.2	0.1	0.4	17.9

Soil temperature. The naked ground without vegetation absorbs a large amount of solar radiation and conducts it, as heat, to varying depths underground. Characteristic of this type of area are urban streets which, in this process, heat up in varying degrees. Figure 23 gives some examples. The extent of heat absorption by the soil depends on heat turnover, that is, on the amount of heat conducted away, on emission from the ground surface, on heat exchange and on the evaporation chill, all of which vary with soil conditions. Constant heat economy values are given in Table 9. The heat conductivity and specific heat of a soil type depend on its water content (Table 10). The heating of the uppermost layer of the soil increases with diminishing conductivity and specific heat and the amount of emission into the air is greater from this layer. The temperature near the ground might surpass that of the surrounding air by 15—30°C with day-time fluctuations of up to 50°C. Intense heating up of naked ground at a positive radiation balance corresponds to severe cooling off at a negative radiation balance.

Air temperature in vegetation areas. As soon as the ground is covered with plants, heat absorption no longer takes place in the soil but in the layer of vegetation. Heat is here consumed by evaporation. The extreme values of heat at ground level and in the uppermost layer of the soil are diminished (Table 11). For heat conditions in a park and in a forest see page 75.

TABLE 8

Attempt to classify some trees according to their photoperiodical characteristics (after Nitsch and others in Lyr et al., 1967)

Species		Country of origin	Type
Acer pseudoplatanus	Sycamore maple	Europe	D ?
Acer rubrum	Red maple	North America	A
Acer saccharum	Sugar maple	North America	B ?
Aesculus hippocastanum	Horse chestnut	Europe	D
Alnus incana	Grey alder	Europe	A
Betula pubescens	Hairy birch	Europe	A
Betula lutea	Yellow birch	North America	A
Betula papyrifera	Paperbark birch	North America	A
Buxus sempervirens	Common box	South Europe	D
Catalpa speciosa	Indian bean	North America	A
Cornus florida	Flowering dogwood	North America	A
Eucalyptus bicostata			
E. niphophila and others	Australian Gum	Australia	C
Fagus grandifolia	American beech	North America	A ?
Fagus sylvatica	European beech	Europe	A+B
Ficus religiosa	Holy tree of Buddha	India	A
Fraxinus americana	White ash	North America	D
Juniperus horizontalis	Creeping juniper	North America	C
Larix decidua	European larch	Europe	A
Liriodendron tulipifera	Tulip tree	North America	A
Morus alba	White mulberry	China	A ?
Paulownia tomentosa	Royal paulownia	China	D
Phellodendron amurense		Asia	A ?
Picea abies	Norway spruce	Europe	B
Pinus sylvestris	Scotch pine	Europe	B
Pinus banksiana and many others	Pines		B
Platanus occidentalis	Plane tree	North America	A
Populus alba	White poplar	Europe	A
Populus nigra	Black poplar	Europe	A
Populus tremula and many others	Poplars		A
Prunus avium	Wild cherry	Asia	D
Pseudotsuga taxifolia	Douglas fir	North America	B
Quercus borealis maxima (Ashe)	Northern red oak	North America	B
Quercus stellata		North America	B
Quercus suber	Cork oak	South Europe	B
Rhododendron catawbiense		North America	B
Rhus typhina	Staghorn sumach	North America	A
Robinia pseudacacia	Locust	North America	A
Syringa vulgaris	Lilac	SE Europe	D
Thuja occidentalis	*Arbor vitae*	North America	C
Thuja plicata		North America	C
Tsuga canadensis	Hemlock	North America	A
Ulmus americana	White elm	North America	A
Viburnum opulus	Guelder rose	Europe	A
Viburnum prunifolium		North America	D
Various tropical woods and *Citrus* species			C

TABLE 9

Some constants for the heat balance of soil (after Geiger, 1961), classified with decreasing heat conductivity

Kind of soil or material	Dense soil		Naturally lying soil			
	Density ρ_s (g cm^{-3})	Specific heat c_s (cal g^{-1} grad^{-1})	Density ρ_m (g cm^{-3})	Volume heat $(\rho c)_m$ (cal cm^{-3} grad^{-1})	Heat conductivity $1000 \cdot \lambda$ (cal cm^{-1} sec^{-1} grad^{-1})	Temperature conductivity $1000 \cdot a$ (cm^2 sec^{-1})
Silver	10.5	0.056	—	0.59	1000 —1000	1700
Iron	7.9	0.105	—	0.82	210	260
Concrete	2.2 — 2.5	0.21	—	0.5	11	20
Rock	2.5 — 2.9	0.17 — 0.20	2.5 — 2.9	0.43 — 0.58	4 — 10	6 — 23
Ice	0.92	0.505	1.7 — 2.3	0.46	5 — 7	11 — 15
Wet sand	2.6	0.20	—	0.2 — 0.6	2 — 6	4 — 10
Wet loam	2.3 — 2.7	0.17 — 0.20	1.7 — 2.2	0.3 — 0.4	2 — 5	6 — 16
Old snow	—	—	0.8	0.37	3 — 5	8 — 14
Water, motionless	1.0	1.0	—	1.0	1.3 — 1.5	1.3 — 1.5
Wet moor	1.4 — 2.0	—	0.8 — 1.0	0.6 — 0.8	0.7 — 1.0	0.9 — 1.5
Loam, dry	2.3 — 2.7	0.17 — 0.20	—	0.1 — 0.4	0.2 — 1.5	0.5 — 2.0
Sand, dry	2.6	0.20	1.4 — 1.7	0.1 — 0.4	0.4 — 0.7	2 — 5
New snow	—	—	0.2	0.09	0.2 — 0.3	2 — 4
Timber, air-dried	1.5	0.27	0.4 — 0.8	0.1 — 0.2	0.2 — 0.5	1 — 5
Dry moor	1.4 — 2.0	—	0.3 — 0.6	0.1 — 0.2	0.1 — 0.3	1 — 3
Air, motionless	0.001 — 0.0014	0.24	—	0.00024 — 0.00034	0.05 — 0.06	150 — 250

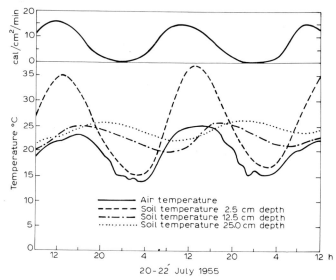

Fig. 23. Daily course of air and soil temperatures at different depths. (After Bonnemann and Röhrig, 1971.)

Temperature of different parts of the plant. The irradiated energy is partly reflected by the leaves and partly transmitted; to a very limited extent it is absorbed. This causes a heating up of the leaves, an effect which is counteracted by CO_2-assimilation but mainly by transpiration. Lastly, heat exchange takes place with the surrounding air. Excess leaf temperatures of 10—20°C are common.

Intense irradiation on a tree mainly strikes the crown surface. Here, too, powerful emission takes place at night, resulting in a cooling off. Within and beneath the tree crown irradiation intensity is reduced (cf. page 37), and emission is screened in the same way. Overheating easily happens in areas exposed to the south or with an incline to the south, as well as in urban streets with strongly reflecting pavements and walls. There it may lead to destruction of the bark (burn) not only in summer, but also in winter. Trees with smooth and thin barks, such as beeches, suffer especially from sunburn.

TABLE 10

Specific heat in cal cm^{-3} from various soils (after Mitscherlich in Walter, 1960)

Kind of soil	Dry	50% saturated	Water-saturated
Sand	0.302	0.510	0.717
Clay	0.240	0.532	0.823
Humus	0.148	0.525	0.902

44

TABLE 11

Heat proportions of uncovered ground and of ground covered with vegetation (after Kirstensen, 1959), in degrees Celsius

Depth (cm)	Uncovered soil			Soil covered with short grass			Diff. 1 less Diff. 2
	Max	Min	Diff. 1	Max	Min	Diff. 2	
2.5	36.6	16.3	20.3	31.9	18.6	13.3	7.0
12.5	25.4	20.1	5.3	24.3	20.9	3.4	1.9
25	23.3	20.6	2.7	22.2	20.9	1.3	1.4
50	19.5	18.6	0.9	19.7	19.1	0.6	0.3
Height above ground level 200 cm				25.0	15.0	10.0	

The heat lifts the bark from the wood; when the temperature drops, the bark does not attach itself closely again. Temperature differences also cause the dreaded frost cracks in the wood of the trunks (see Fig. 24 and page 194).

The higher the transpiration rate of trees, the lower the temperature fall,

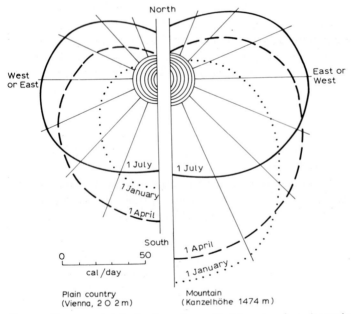

Fig. 24. Heat sum-total of a day of a vertical tree trunk on its various sides on January 1st, April 1st and July 1st (after Krenn, 1933). The length of the radial lines up to the point of intersection with the curves shows the amount of heat in cal per day at the analogous exposition.

but this is possible only when a tree is well supplied with water. Then the cooling which results from transpiration reaches high values at conditions of high air temperature and low air humidity. If a further transpiration is brought about by wind (within certain limits), the leaves may be a few degrees cooler than the surrounding air. Treatment with antidesiccants (antitranspirants) promptly raises leaf temperatures. Plant reactions to the heat factor vary because individual processes such as assimilation, respiration, growth etc. reach their maxima at different temperatures. What is known as the phenological calendar, which shows when various plants start flowering, also makes clear how they are affected by heat. In this way maps of whole countries or geographical areas can be drawn showing the start of spring. For every one degree increase of latitude (111 km) spring will start four days later. Plants can thus be regarded as the best meteorological stations. In phenological terms, urban areas are far ahead of their environment.

The importance of heat for the vital processes of trees. Like other plants, trees also have optimum temperatures for their vital processes. Deviations upwards or downwards are most disturbing. As a rule the production of matter and the greening of woody plants begins at an average daily temperature of $10°C$. Termination of assimilation activity and the discolouring of leaves can be observed at similar average daily temperatures (Giessen/Western Germany 8—9°C, Leningrad/Russia 10°C, Moscow 13.8°C). Northern trees (e.g. spruce) have modest requirements, southern trees are very demanding. The longitudinal growth of trees is more affected by night-time than by day-time temperatures (Larcher, 1973a). Kozlowski (1971a) cites Mikola (1962) who says that longitudinal growth is regulated by temperatures of the previous summer, as a result of carbohydrate reserves. At any rate, it is to be noted that growth terminates before temperature conditions compel such termination. Therefore heat is not the growth-inhibiting factor. Obviously root growth is also regulated by temperature variations.

Effect of extreme temperatures. The overheating of plants is counteracted by transpiration, which causes land plants to consume up to 99% of the absorbed irradiation (Lyr et al., 1967). If critical threshold values are surpassed, cells die of heat or cold (Tables 12 and 13). For more on this subject see Larcher (1973a,b). Statements concerning the frost-hardiness of trees must be carefully considered, since this usually depends on a series of related factors. It also makes a difference whether the frost occurs at the beginning or near the end of the vegetative period or during the dormant season.

4.1.3. Carbon economy

Photosynthesis transforms the absorbed radiation energy into chemical energy, which is stored. At the same time sugar is produced from inorganic matter. Details may be found in publications dealing especially with this sub-

TABLE 12

Maximum temperature resistance of leaves and buds of autotrophic terrestrial plants.
(Data from numerous authors according to Larcher, 1973b)

Plant distribution	Low-temperature injury in winter (cooling 2 hours and more) (°C)		Heat injury in summer (heating 30 min) (°C)
	Leaves	Buds	Leaves
Tropical regions			
Evergreen rain forest			
Ferns, herbaceous angio-sperms	+5 to −2		+45 to +48
Mangroves	+5 to −4		about +50
Drought deciduous woodlands	−3 to −7	−5 to −8	+45 to +55
Subtropical dry regions			
Succulents and C_4-plants			+50 to +60
Species with insufficient transpiration cooling			+50 to +57
Species with efficient transpiration cooling	−8 to −12		+45 to +48
Warm temperate zone			
Mediterranean hard-leaved forests	−6 to −13	−8 to −16	+50 to +55
Warm temperate forests of coasts and islands	−7 to −12	−12 to −15	
Temperature zone			
Dwarf shrubs of Atlantic heaths	about −20	about −20	+45 to +50
Submediterranean woody plants	about −20	−25 to −30	about +50
Trees and shrubs of wide distribution in the temperate zone	−25 to −35	−25 to −40	about +50
Winter-cold regions			
Alpine dwarf shrubs	−20 to −70	−20 to −50	+47 to +54
Evergreen conifers	−40 and below	−35 and below	+44 to +50
Deciduous boreal trees and shrubs	−40 and below	−40 and below	+45 to +48

TABLE 13

Frost resistance (temperature T_i ($^\circ$C) at the first occurrence of injuries), frost avoidance (as essentially limited by the tissue-freezing temperature T_f ($^\circ$C)), and frost tolerance (expressed as difference between T_i and T_f) of evergreen leaves in winter. (Data from several authors according to Larcher, 1973b)

Species	Resistance (T_i)	Avoidance (T_f)	Tolerance $(T_i - T_f)$
Eucalyptus globulus	—3	—3	nil
Citrus limon	—5	—5	nil
Ceratonia siliquata	—5	—5	nil
Nerium oleander	—7	—7	nil
Olea europaea	—10	—10	nil
Arbutus unedo	—9	—6	3
Pinus pinea	—11	—7	4
Quercus ilex	—13	—8	5
Cupressus sempervivum	—14	—5	9
Cedrus deodora	—15	—6	9
Taxus baccata	—20	—6	14
Abies alba	—30	—7	23
Picea abies	—38	—7	31
Pinus cembra	—42	—7	35

ject. The exchange of carbon dioxide and oxygen between cells and their surroundings occurs by diffusion. This in turn is regulated by the stomata which also control transpiration. The extent of this diffusion, as well as the quantity of gaseous exchange between cells and environment, depends on the number of stomata per mm^2 of leaf surface (Table 14). It should be noted that despite high density of stomata the percentage area of pores on the whole of the leaf surface is very small. This is especially true of the evergreen hard leaf plats of the macchia.

TABLE 14

Stomatal frequency, width and area of pores of different groups of plants (after several authors in Larcher, 1973a, shortened

Plants	Stomatal frequency/mm^2	Stomatal length (μm)	Stomatal width (μm)	Area of pores in % of leaf surface
Grass	(30) 50 — 100	20 — 30	about 3	0.5 — 0.7
Deciduous trees	100 — 500	7 — 15	1 — 6	0.5 — 1.2
Evergreen coarse-leaf plants	100 — 300	10 — 15	1 — 2	0.2 — 0.5
Conifers	40 — 120	15 — 20	—	0.3 — 1

The capacity of the stomata to open and, with it, diffusion resistance depend on environmental factors, namely: light, supply of carbon dioxide, temperature, air humidity, supply of water; and internal factors, namely: partial CO_2 pressure, water condition of the plant, ionic cycle, and phytohormones.

Light. The brighter the light, the wider the stomata open, provided there is adequate water supply ("photo-active opening" according to Stolfeld), but only about one third of the actual radiation is needed in this process (in tropical rain-forests about one half). Light and shade leaves possess different points of compensation in relation to the CO_2 gas exchange.

Carbon dioxide supply. The amount of CO_2 available in the air surrounding the leaves is relatively small. In calm, unmoved air the amount is lowered as a result of photosynthesis. A light breeze (up to 1.7 m/sec) may bring along carbon dioxide and increase its uptake and assimilation. However, this rise of assimilation with increasing CO_2-supply is regulated by light and temperature.

Temperature. The stomata open more rapidly with rising temperatures. Lower temperatures, e.g. below 5°C, slow down the movement of the guard cells. Below zero, down to −5°C they stay locked. In poor light conditions plants assimilate more at low temperatures than at high ones. Strong light intensity has the opposite effect. Above all, high temperatures intensify respiration.

Conifers become dormant in winter only after fairly long periods of cold weather, and then show no gaseous exchange even at temperatures a little above zero. Generally, evergreen plants remain passive or very cold days, but start their production again as soon as the days get warmer. High temperatures stop the dark-reaction in photosynthesis by making the enzymes inactive and by disturbing the reciprocal effects of the various reactions. The optimum temperature for trees is somewhere between 20 and 30°C. Cold and heat limits for carbon dioxide uptake are given in Table 15.

Water supply. This plays an important role in the behaviour pattern of stomata. With decreasing turgor they close within 10—15 minutes. Only in a narrow range of optimum water supply is the uptake of CO_2 high. The complete range of reactions of stomata is more fully explained in the chapter on water economy. *Cryptomeria* and *Pinus densiflora* assimilate even when water saturation in the soil drops to 25% (Polster in Lyr et al., 1967).

Nutritive condition. Good nutritive condition of the trees stimulates photosynthesis in two different ways: it increases assimilation and stimulates

TABLE 15

Temperature dependence of net-photosynthesis during one vegetative period with natural CO_2 supply and light saturation (after investigations of numerous authors in Larcher, 1973a, shortened)

Group of plants	Cold limit of CO_2 uptake (temperature minimum of Ph_n) ($^\circ C$)	Temperature optimum of Ph_n	Heat limit of CO_2 uptake (temperature maximum of Ph_n)
Agricultural food plants (C_3)	−2 to 0 and more	20 − 30 (40)	40 − 50
Woody plants			
Tropical and subtropical evergreen broad-leaved trees	0 to 5	35 − 30	45 − 50
Coarse-leaved shrubs and trees from dry zones	−5 to −1	15 − 35	42 − 55
Deciduous trees of the temperature zone	−3 to −1	15 − 25	40 − 45
Evergreen conifers	−5 to −3	10 − 25	35 − 42

leaf growth. The mineral salts are building blocks for chlorophyll, enzymes, pigments and for the activators of photosynthesis. On the other hand, a low nitrogen level combined with strong doses of potassium may impede photosynthesis (Polster in Lyr et al., 1967).

Interaction of external factors. The external factors listed influence the organism and regulate the width of the stomata which are mostly medium-sized, since the positive factors seldom all operate at once (Table 16). In Central Europe the most marked assimilation occurs in cool weather conditions coupled with an amply supply of light. Shortly after sprouting the young leaves assimilate but little. Later, when the leaves are fully grown, photosynthesis increases rapidly until another decline takes place in the autumn. Conifers have a high point of compensation, which means that assimilation exceeds respiration only at a late stage. Polster (in Lyr et al., 1967) sets the diminution of CO_2 uptake at 20—30%, and the highest daily values of photosynthetic activity at 70—80%. In medium latitudes the lack of light limits photosynthesis while temperatures there do not greatly affect deciduous trees. But cold weather in northern latitudes and heat in southern ones hinder assimilation. However, the strongest inhibiting factor in all cases is water deficiency.

50

TABLE 16

Diminution of the CO_2-uptake due to production-limiting factors (after Schulze, 1970) on a mature beech in West Germany

Factor	Average diminution during the vegetative period (%)	Maximum dimuntion during unfavourable weather periods (%)
Lack of light at dawn and dusk	−22	
Lack of light from clouds and fog	−16	−56
Unfavourable temperatures	− 3	−15
Air aridity	− 2	−13
Total diminution	−43	
Remaining efficiency	+57	

Extent of net photosynthesis. The maximum hourly CO_2 uptake amounts to 150 mg/dm² of leaf surface under optimum conditions in terms of light, water supply, temperature and artificial increase of CO_2 content in the air to 1 vol. % (Larcher, 1973a). In free natural conditions, however, CO_2 absorption at best amounts to only 80 mg/dm²/hour. Conifer needles make poor use of the light so that overall conifer assimilation stays below that of deciduous trees, but on account of their large leaf area index they equal and sometimes even surpass their deciduous brothers.

Respiration. The respiration intensity of trees, as well as of other plants, varies with developmental stages, growing activity and temperature. The dark respiration of mature leaves in summer (temperature 20°C) runs up to 3—4 mg g^{-1} dw h^{-1} with light leaves of deciduous trees (shade leaves 1—2), and with evergreen confiners is about 1 mg g^{-1} dw h^{-1}. This is about half of that with wild herbaceous plants (Larcher, 1973a). Tree respiration is at its peak shortly before sprouting of the leaves and again before shedding. Shoot growth in warm spring weathers so intensifies the respiration of spruce, douglas fir and pine that the needle assimilates are unable to compensate for it. This also happens after an autumn frost (Polster in Lyr et al., 1967). The minimum temperature required for respiration is between −10 and −25°C. Rising temperatures in summer increase respiratory intensity. The tree trunk, its bark and the outermost layer of cambium cells respire especially intensively, and the roots more so than the shoot axes. Temperatures around 25°C can cause such high respiratory intensity in light wood species that up to half of the assimilates produced is used up again in respiration (Polster, 1950).

Production of dry matter. The amount of carbon in the dry matter of trees is increased by the surplus of the CO_2 balance. In its first years of life, the wooden substance and the leaf substance of a tree are about equal. Mature deciduous trees in the temperate zones have only a 1—5% share of photosynthetically active organs, and evergreen conifers in Taiga and mountain forests 4—5%, so that this foliage is forced to nourish 20—100 times its own mass of matter and deliver the necessary building substances and nutrients to it (Larcher, 1973a). This again shows how necessary it is, in view of the severe losses a tree may suffer, to consider every pruning of trees with the greatest care.

The distribution of assimilates inside the tree is a very complicated process. Shortly before the leaves sprout in spring assimilates stored during the previous year in trunk, bark, roots and branches are transported to the buds and new shoots. They serve as nourishment for young shoots and leaves which, when mature, in turn start producing new assimilates for leaves, shoots, flowers, fruits and then for the cambium and winter buds. Finally the supply in the roots and bark is replenished. The surplus goes to next year's flower buds. Sometimes this is very little, which is why many temperate-zone trees bear fruit only at intervals of several years. During the dormant period assimilates are again stored in the wooden body of the tree. In spring evergreen trees still have the assimilatory organs of the previous year, which start to assimilate very early and so provide for the new growth. Such trees are thus at an advantage in places where the growing season is shortened by a long winter or by a long period of summer drought.

For details concerning the total production of matter by trees see pages 23 and 160 onwards. Please note, however, that all figures are approximate. The term "Net Primary Production" (NPP) used in current ecological literature denotes the total amount of dry matter produced by a plant.

Oxygen economy. This is closely connected with the carbon economy. For details see pages 23 and 160 onwards. Oxygen consumption in the soil by the tree roots is very high, and even under favorable conditions the supply is replenished only very slowly. For details see page 59.

4.1.4. Water economy

Water is not a nutrient in the ordinary sense. Rather, it serves to condition the plasma and so enables it to carry out the processes of life. Walter (1960) calls this "hydrature" (Hydratur). It will reach its highest effect when water is freely available, that is, when no osmotically acting substances have been dissolved in it.

Water absorption from the whole of the plant's surface. Usually trees can absorb water on the whole of their surface. Whether the quantities thus absorbed are of importance is another question, however. Dew precipitation

TABLE 17

Dew quantities in various localities (after Walter, 1960)

Author	Locality	Dew quantity
Leick	Greifswald North Germany	20% of yearly precipitation
Hittner	München	10% of yearly precipitation
Bernick	Hiddensee/Ruegen Germany	7% of monthly precipitation June through September
Walter	Virginia/U.S.A.	6.9% of precipitation of ten months
Walter	Desert South Africa	40 mm/year

is generally very limited (Table 17). The ecological evaluation of dew precipitation is not concerned with total amounts but with precipitation per night, which in Central Europe is below 1 mm. Walter cites Hoffmann who gives a maximum of 0.07 mm/hour. Often water released by guttation is taken for dew. With a few exceptions (Table Mountain in Capetown, fog valleys in California with *Sequoia sempervirens*) dew precipitation cannot improve the water balance. Dew is of greater importance, however, for transpiration reduction, particularly in the case of a long-lasting morning fog. On critical days this aids the water balance, especially for shallow-rooted trees. On the other hand, dew and fog favour fungus infections. Of course rain can also be absorbed by the leaves, but this kind of water uptake by the whole of the plant's surface is of small significance. Only in times of emergency can it be a help.

Water condensation in the soil. If the soil temperature is considerably lower than the air temperature the water vapour of the warm air penetrating into the soil is able to condense. Trenel (1954) suggests a diffusion of the molecules of water vapour. As a result of the heating of the uppermost moist layers of soil by irradiation, the vapour pressure increases. Thus moisture from dew may penetrate into deeper soil layers as a result of the drop in vapour pressure. Excavations near street trees in Berlin (Hoffmann, 1954) have shown that the undersides of the flagstone pavement above the root areas of the trees were quite moist and many fine fibrous roots were found there. These phenomena and their interrelation await further clarification.

Water absorption from the soil. Part of the precipitation goes down into deep soil layers as "sinking water". Another part is stored in soil cavities as "adherent water". There is no precise borderline between them. The higher the water capacity (= quantity of water the soil is able to hold against the force of gravity) the more adherent water is held back after rain-showers,

TABLE 18

Water bound by hygroscopic force to various kinds of soil (after Klemm in Lyr et al., 1967)

Kind of soil	Water bound hygroscopically (%)
Sand	1
Loam	3 — 6
Clay	10 — 12
Humus	18 — 26

and the less deeply will the ground be moistened. Part of the water is bound hygroscopically with more than 50 atm to the surface of the soil particles (Table 18). The absorptive capacity of the roots is not strong enough to get hold of this water. The adherent water in the capilaary cavities of the soil is held there with a pressure of only 0.33 atm; it is therefore more easily available to the roots which develop absorptive capacities of 5—15 atm. Generally the osmotic values of the roots exceed those of the soil solutions by about 5—6 atm. The capillary rising capacity of the water is mostly overestimated. It largely depends on soil conditions. In loamy soil the zone of the enclosed capillary water above ground water level amounts to only 40 cm, adjoined by a 1.10 m-thick layer of rising film-water. Thus the total rising capacity amounts to only 1.50 m. In pure sand both values decline to a little above 30 cm (Walter, 1960). The rising speed of the water is moderate indeed.

Evidently water absorption in the soil takes place by a combination of osmotic energy and active water transport caused by the processes of metabolism. The roots as organs of water absorption continuously grow towards moist layers of soil and "graze" water from the soil particles as long as their suction tension is greater than that of the soil. Low soil temperatures reduce water absorption as a result of decreasing plasma permeability. They also slow down root growth. Extreme concentrations of CO_2 and lack of oxygen further diminish plasma permeability.

Water transport. The transport of water up to the crown proceeds by suctional tension from below upwards to the leaves. Lyr and Klemm (in Lyr et al., 1967) give a good illustration of the factors affecting water absorption and transport (Fig. 25). The speed of water transport and the transmitting capacity of the vascular pipeline depend on the potential decline from leaves to roots and on conductivity resistance, hence on the structure of the whole system (Table 19). *Fraximus, Castanea, Quercus* and *Robinia* are ring-pored. *Acer, Carpinus, Fagus, Juglans* and *Populus* have dispersed pores. The specific conductive capacity of conifers is half of that of broadleaved evergreens, and this again is half that of deciduous woods.

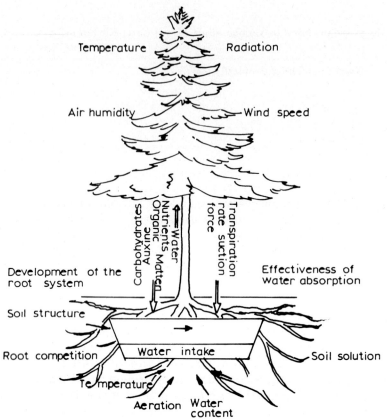

Fig. 25. Schematic illustration of the factors controlling absorption of water (after Lyr and Klemm in Lyr et al., 1967).

Transpiration. To a small extent transpiration proceeeds via the cuticle (normally less than 10% of free evaporation) and to a larger extent through the stomata. The gaseous exchange necessary for assimilation via the open stomata is also the engine for stomatal evaporation. The latter therefore occurs only on account of assimilation. Possibly the transport of nutrients from the soil into the leaves likewise depends either on transpiration of the cuticle or on guttation (Table 20). The stomatal diffusion resistance in trees, especially conifers, is far greater than in herbaceous plants. Also to be noted is the difference in stomatal frequency. Transpiration is always related to carbon dioxide absorption (page 48), as well as to irradiation, temperature and the aridity and flow of air. The two cycles regulating the movements of the stomata under the influence of carbon dioxide and water have already been mentioned (page 48). Even at low water stress trees reduce the stomata aperture. Deciduous trees raise their transpiration intensity much

TABLE 19

Specific axial conductivity of the xylem and maximum speed of the transpiration flow (after Huber, 1956)

A. Specific conductivity of the wooden body $\left(cm\ h^{-1} \cdot \dfrac{atm}{m}\right)$

Conifers	20
Broad-leaved evergreens	13.5 — 48
Deciduous trees	65 — 128
Lianas	236 — 1273
Root-wood of deciduous trees	292 — 5388

B. Maximum speed of the transpiration flow ($m \cdot h^{-1}$)

Conifers	1.2
Larch	1.4
Mediterranean hard-leaf plants	0.4 — 1.5
Deciduous trees with dispersed pores	1 — 6
Deciduous trees with ringed pores	4 — 44
Herbaceous plants	10 — 60
Lianas	150

more quickly than do conifers with their needles. The larch occupies an intermediate position. Only pines keep transpiration going on when other species of trees have long since reduced it.

On the shady sides of trees the foliage evaporates 1/4 less water than on the sunny side. A solitary tree exposed to the wind consumes more water than does a tree in a forest. Larcher estimates the total evaporation of a

TABLE 20

Transpiration of leaves of various trees with evaporation capacity of the air of 0.48 ml $H_2O \cdot h^{-1}$. All data given in mg $H_2O \cdot dm_2^{-2} \cdot h^{-1}$ (after Pisek and Cartellieri, 1931, 1932, 1933; Pisek and Berger, 1938 in Larcher, 1973a, shortened)

Species	Total transpiration at open stomata	Cuticular transpiration after closing of stomata	Cuticular transpiration in % of total amount
Betula pendula	780	95	12
Fagus sylvatica	420	90	21
Picea abies	480	15	3
Pinus sylvestris	540	13	2.5

56

TABLE 21

Average water consumption in the process of dry-matter production in g of water per g of produced dry matter (after Stocker, 1929; Polster, 1967; Black, 1971 in Larcher, 1973a, shortened)

Oak	340	Pine	300
Birch	320	Larch	260
Beech	170	Spruce	230
		Douglas fir	170

For comparison: Rice 680 Wheat 540

birch at 100 l a day, that of a beech at 9000 l per vegetative period (i.e., 50 or 75 l of water daily, depending on whether the length of the vegetation period is 180 or 120 days). Moderately opened stomata offer the best compromise between water evaporation and the uptake of carbon dioxide.

The relation between water consumption and dry-matter production is measured as the transpiration coefficient. Values are shown in Table 21. Questions concerning the specific transpiration of different trees species can only be answered with reservations. Conditions in the open country differ greatly from those prevailing in the laboratory, where measurements are made, so that comparisons and generalizations of the results obtained by

TABLE 22

Daily transpiration of trees and shrubs in g water per g of fresh foliage or needle mass according to different authors (after Lyr and Klemm in Lyr et al., 1967)

Tree species		Transpiration	Author
Populus alba	White poplar	13 — 14	Stocker, 1929
Elaeagnus angustif.		13 — 14	Stocker, 1929
Populus nigra sect. Aigeiros	Black poplar	9 — 15	Polster, 1957
Betula pendula	Silver birch	8.1	Pisek and Cartellieri, 1939
Quercus robur	English Oak	6.0	Polster, 1950, 1957
Corylus avellana	Hazel	4.2	Pisek and Cartellieri, 1939
Fagus sylvatica	Beech	3.9	Polster, 1950, 1957
Larix decidua	Larch	3.8	Pisek and Cartellieri, 1939
Pinus cembra	Swiss pine	2.2	Berger-Landefeldt, 1948
Pinus strobus	White pine	2.1	Pisek and Cartellieri, 1939
Pinus sylvestris	Scotch pine	2.0	Pisek and Cartellieri, 1939
Picea abies	Norway spruce	1.4	Polster, 1950, 1957
Pseudotsuga menziesii	Douglas fir	1.3	Polster, 1950
For comparison: Oxalis acetosella	Woodsorrel	1.5 — 2.0	Berger-Landefeldt, 1948

TABLE 23

Transpiration of different species of deciduous tree and conifers given in g of needle or leaf fresh weight (after Bonnemann and Röhrig, 1971)

Tree species	Daily transpiration in g fresh weight of needles and leaves according to		
	Polster	Schubert	Pisek and Cartellieri
Spruce	1.4	1.1	1.4
Pine	1.9	1.9	2.0
Douglas fir	1.3	—	—
Larch	3.3	4.1	3.8
Oak	6.0	—	7.6
Beech	4.8	—	2.8
Birch	9.5	—	8.0

different authors cannot safely be made. Only absolutely equal locations would permit genuine comparison (Tables 22 and 23).

The highest water deficit, without damage, after the closing of stomata can be endured by *Pinus sylvestris*, whereas beeches can endure only the lowest. On the shaded sides of a spruce crown standing in water-saturated soil the aridity of the air on warm days suffices to induce the stomata to close. Daily transpiration values, already unreliable in many respects, become even more uncertain when converted into yearly values or applied to whole forests (cf. page 76 onwards).

Frost dryness. Water deficit results when frost blocks its further supply. Below −2°C water freezes in the trachea and tracheides while the branches and trunk continue to lose water. This happens particularly in late winter when the sun has already begun to stimulate transpiration. To avoid frost dryness deciduous trees shed their leaves in the autumn, but in the much colder North the conifers keep their needles! The reason is that they are better protected against water losses. Some deciduous trees of northern regions also show smaller transpiration losses than do similar species in southern areas. The extent of frost penetration into the soil depends on its water content and on its capacity to conduct heat.

Resistance to aridity. Drought resistance takes the form of either drought avoidance or drought tolerance. The following measures are necessary for drought avoidance: improvement of water absorption from the soil by expanding the root area and increasing suctional capacity; reduction of water output by closing the stomata, reducing transpiratory surfaces and cuticle protection; water storage and increased water transport capacity. The ability to survive is a measure of aridity resistance. It shows how long the leaves of a tree with closed stomata and at a given evaporation rate, will remain undamaged. The survival figure for *Pinus sylvestris* is 50 hours; for *Picea abies* 22 hours; for *Fagus silvatica* 2.1 (shade leaves) or 0.7 (light leaves) hours; for *Quercus robur* 1.5 hours (Larcher 1973a, further details there).

Water balance. The water balance shows the value of water uptake by absorption minus water output by transpiration. The distance travelled between uptake by the roots and output by the leaves is very considerable, so that trees must guard especially against a negative water balance. They do this by reducing transpiration at noon. They may thus be called hydrostable (isohydric), though not all to the same degree. Transpiration reduction begins in the shade leaves, moves on to the base of the crown and finally reaches the top leaves. The water-conducting parts of the trunk, branches and the large roots serve as water storage places, from where temporary alterations in the water balance may be leveled out. Conifers possess a very even water balance, as do species of shade-wood trees and many light-wood species such as oaks.

4.1.5. Mineral economy
The soil. Soil is the basic source of nutrition for trees and all other plants. Soil science deals with its origin, structure and dynamics (Scheffer and Schachtschabel, 1973). The formation of soil is a complex process in which trees as well as other plants take part (Fig. 26). The process of soil development is essentially the breaking up of primary clay minerals and the formation of secondary clay colloids. This process involves a continuous recurrence of disintegration followed by a delivery of nutritive substances so that, theoretically, trees never need to be fertilized. Besides inorganic components the soil also contains organic sustances. The latter comprise
the edaphic living flora and fauna of the soil (edaphon),
dead organic matter (humus).

The organic substance is essential for the life of soil organisms. It also improves the soil's physical condition. Humus binds the nutritive substances, thus saving them from being washed deeper into the ground. It also improves the water retaining capacity and the aeration of the soil. One gram of sloughy loam contains (after Burrichter in Scheffer and Schachtschabel, 1973)

in meadows	24.8 billion bacteria,
in fields	13 billion bacteria,
in the forest floor	10.1 billion bacteria.

In addition the soil harbours fungi, algae and vast numbers of animals from protozoa to metazoa and earthworms. The weight of the total microflora, microfauna and metazoic fauna in the uppermost 15 cm of agriculturally used soil amounts to ca. 25 t/ha (Stoeckli in Scheffer and Schachtschabel, 1973). The moisture requirements of the edaphon vary widely. Most favourable for bacterial life is a moisture content of 60—80% of the soil's field capacity. The optimum moisture content for fungi is lower than that for

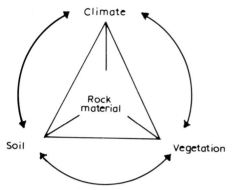

Fig. 26. Process of soil development.

actinomyceta and bacteria. Thus the soil's fungi content is increased in relation to decreasing water content. A high water level displaces the soil air and with it the oxygen, but as all of the soil fauna and most of its flora require aerobic conditions, the lack of oxygen produces a negative effect. Most soil organisms possess an optimum temperature between 25—35°C. Temperatures above 80°C kill the majority of soil organisms. Bacteria and fungi are highly cold-resistant.

Soil reaction. This, too, is an important factor in terms of the tree's position, but it should not be over-stressed as soil reaction is related to general soil conditions such as aeration, water supply, etc. The pH value alone does not tell us a great deal. Fungi also tolerate an acid soil reaction. The optimum growth of bacteria and actinomyceta requires neutral to weak alkaline reactions. Pirone (1972) gives soil reaction adaptations of some trees.

Soil aeration. Generally the soil's air content at field capacity is

in sandy soil	30—40%,
in loamy soil	10—25%,
in clay soil	5—15%.

However, in the open country the soil's air content is usually higher because it rarely happens that the soil is water saturated; thus considerably more soil pores contain air.

Respiration by plant roots and soil organisms diminishes the oxygen content of the soil air and increases carbon dioxide. Soil air contains an average of 0.2—0.7% CO_2 and over 20.3% oxygen. With increasing depth the carbon dioxide content increases while the oxygen content decreases. When the latter sinks below 18—12% the roots are damaged. At less than 11% they die (Ruge, 1972c). The average amount of soil-borne carbon dioxide is estimated at 1200 kg/ha/year by Scheffer and Schachtschabel (1973); one-third of this amount derives from root activity, two-thirds from microorganisms.

The more coarse pores a soil contains the quicker the gas exchange takes place. The moister and thus the more compact the soil, the more aeration is impeded. Lack of air impedes the biological activity of microorganisms as well as root growth and the uptake of water and nutrients. The soil air is closely related to soil water, since with increasing water content the air content is decreased.

Tree root survival under flood conditions. Waterlogged soils damage or kill roots due to the lack of oxygen and the intensified production of ethanol, which causes the plant to poison itself. Experience has shown that flood-tolerant trees exist; this prompted investigations by Chirkova (1968), Gill (1970), Vester (1972) and Crawford (1974). Some plants have active oxygen diffusion from shoot to root; however, trunk development effectively impedes this diffusion. Chirkova ascertained that the lenticels of the genus *Salix* play a role in permitting the escape of volatile toxic substances and in the downward diffusion of oxygen, but this is probably effective only at short distances. Vester explains flood-tolerance as resulting from morphological adaptations facilitating gas exchange between root and shoot, and from metabolic properties. Trees which can develop new roots at ground surface level are flood-tolerant. However, oxygen supply is not the only factor enabling trees to survive. In most flood-tolerant plants alcohol is the usual product of anaerobiosis. When flooded, these plants steadily increase their rate of ethanol production. However, in flood-tolerant trees there are a large number of substances which can accumulate during anaerobiosis without any toxic effects on the plant cells. These include malic, lactic and shikimic acids, glycerol and certain amino acids. On return to air after periods of partial anoxia, the oxygen debt accumulated in the form of anaerobic respiration products has to be repaid (Vester, 1972; Crawford, 1974).

Flood tolerance may be linked to the production of certain metabolites in the roots and by the translocation of anaerobic respiration products from the roots to the aerial sections of the trees. Differences in the $C^{14}O_2$ fixation of the roots play a role.

A higher root/shoot ratio leads to greater flood tolerance (greater capability to withstand waterlogged conditions). Crawford gives a list of intolerant tree species and for each group of intolerant trees he suggests a more flood tolerant species, which bears some ecological or taxonomic relationship with the intolerant tree (Table 24). The flood tolerance of the trees mentioned is not absolute, but subject to time limitations. According to Crawford it may also be assumed to result from lack of oxygen caused by concrete, traffic and gas leakage.

Mineral substances and soil. Less than 0.2% of the mineral substances in soil is dissolved in the water. Almost 98% is bound to organic waste, to humus or inorganic compounds, or incorporated into minerals. The remain-

TABLE 24

Tree species intolerant to flooding with suggested replacements from taxonomically related groups which are known to withstand flooding (Crawford, 1974) and suggestions from other authors

Kind	Intolerant	Tolerant
Beech	*Fagus sylvatica*	*Nothofagus dombeyii* *N. antarctica* *N. pumilo*
Elm	*Ulmus glabra* *U. procera* *U. carpinifolia*	*Ulmus americana* *U. alata* *Celtis occidentalis*
Ash	*Fraxinus excelsior*	*Fraxinus pennsylvanica* *F. chinensis*
Sycamore and maples	*Acer pseudoplatanus* *A. campestre* *A. platanoides*	*Acer saccharinum* *Platanus* x hybrida *P. occidentalis*
Holly	*Ilex aquifolium*	*Ilex decidua*
Oak	*Quercus robur*	*Quercis petraea* *Q. palustris* *Q. phellos* *Q. shumardii*
Eucalypts and myrtles		*Myrceugenella apiculata* *Myrceugenia exsucca*
Locusts		*Gleditsia triacanthos*
Pine	*Pinus*	*Pinus contorta* *P. thunbergii* *P. taeda* *P. palustris*
Larch	*Larix decidua*	*Larix laricina* *Taxodium distichum* *T. ascendens*
Cedar	*Cedrus libanotica* *C. deodora* *C. atlantica*	*Libocedrus chilensis* *Fitzroya cupressoides*

Author		
Polster (in Lyr et al., 1967)	*Celtis occidentalis* *C. laevigata* *Liquidambar styraciflua* *Ulmus americana*	*Populus* *Salix* *Alnus* *Fraxinus profunda* *Nyssa aquatica* *Prunus padus*

TABLE 24 (continued)

Author	Intolerant	Tolerant
Kruessmann, 1974	*Acer saccharum*	*Acer rubrum*
	Betula papyrifera	*Malus* 'Dolgo'
	B. populifolia	*Morus alba*
	Cercis canadensis	*Fraxinus americana*
	Cladastris lutea	*Juglans nigra*
	Cornus florida	*Salix alba*
	Crataegus lavallei	*S. discolor*
	Magnolia soulangiana	*Tilia cordata*
	Malus species	
	Prunus persica	
	P. serotina	
	P. subhirtella	
	Quercus rubra	
	Robinia pseudacacia	
	Sorbus aucuparia	
	Picea abies	
	P. pungens	
	P. pungens 'Glauca'	
	Taxus cuspidata 'Expansa'	
	T. media 'Hicksii'	
	Thuja occidentalis	
	Tsuga canadensis	

ing 2% is absorbed by soil colloids. They are taken up in the form of ions and built into the plant mass or stored in the cell sap.

The tree takes nutritive substances out of the soil by
absorption of nutritive ions out of the soil water,
absorption exchange of absorbed nutritive ions,
disintegration of bound nutritive stores (Larcher, 1973a).
The roots absorb nutritive ions selectively but they cannot avoid noxious salts (de-icing salt).

Mineral substances are taken in from under the ground (decomposition), from fertilization (page 264) and from water and air. In Europe, the supply of mineral substances from the air amounts to about 25—75 kg/ha/year; mainly N, Cl, Na, S, K and Mg.

The transport of the mineral substances in the tree takes place in the xylem up to the final points of the vessel netting. In the joints of roots and shoots the conducting lines of xylem and phloem are connected so that in the flow of phloem, as well, mineral substances can be brought to the points of need. For many trees, the leaves shed return to the soil the larger part of the absorbed mineral substances. In a spruce forest in Western Germany these were found to be 93% of the potassium, more than 80% of the calcium and magnesium and more than 70% of the absorbed nitrogen and phos-

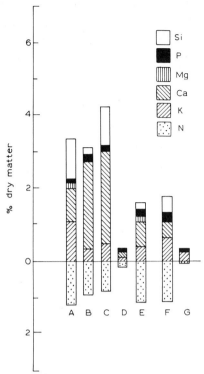

Fig. 27. Concentration and distribution of minerals in various organs of trees (Ehwaldt, 1957).
Beech: A = leaves; B = twigs; C = bark; D = mature wood; E = fibrous roots.
Spruce: F = needles; G = trunk.

phorus (Larcher, 1973a). The removal of all the leaves, twigs and other litter in municipal parks and in the streets entails the loss of these mineral substances for the ecosystem. For deciduous trees and conifers the average ash content of dry matter amounts to about 3—4% in the leaves and needles respectively; in the wood it is about 0.5% and in the bark 3—4 (8%) (Fig. 27).

4.1.6. Nitrogen economy

Quantitatively, nitrogen ranks fourth among the bioelements. The nitrogen economy is closely connected with the carbon economy and its energy delivery, just as the production of matter depends on the supply of nitrogen. Poorly aerated soils are inadequately provided with nitrogen. Although the main supply of nitrogen is to be found in the air in a gaseous state, the plant absorbs it in ionic form out of the soil where nitrogen is fixed by free and symbiotically living bacteria. Photosynthesis and good soil aeration are basic requirements for the binding of nitrogen. Trees and shrubs capable of bind-

64

TABLE 25

Trees and shrubs able to bind atmospheric nitrogen (after several authors)

Leguminosae

Amorpha	*Laburnum*
Caragana	*Robinia*
Colutea	*Sarothamnus*
Cytisus	

Gleditsia, Gymnocladus and *Sophora* are not capable of binding nitrogen.

Non-*Leguminosae*

Alnus	*Elaeagnus angustifolia*
Casuarina species	*Hippophae rhamnoides*
Ceanothus species	*Myrica gale*
Comptonea peregrina	*Shepherdia* species

TABLE 26

Depth distribution of fibrous roots and nodules of unshaded and shaded *Robinia* trees in the root cellar Eberswalde on filled-in sandy soil; data per plant and per 1 cm layer of soil (after Hoffmann, 1960 in Lyr et al., 1967)

Soil depth	Unshaded (full daylight)			Shaded (30% of full daylight)		
	Fibrous roots (mg)	Nodules		Fibrous roots (mg)	Nodules	
		Number	Weight (mg)		Number	Weight (mg)
5	180.0	52.0	148.9	12.4	1.6	2.5
15	117.8	17.2	37.8	18.0	4.0	1.6
35	85.6	9.1	39.0	5.8	2.7	0.9
55	21.1	1.1	13.3	4.8	0	0
75	46.7	0.3	3.1	7.9	0	0
95	40.0	0.2	5.6	6.2	0	0
115	42.2	0.1	1.4	9.2	0	0
135	14.4	0	0	2.7	0	0
155	18.9	0	0	4.2	0	0
175	38.9	0	0	7.3	0	0
195	40.0	0	0	2.1	0	0
205	18.3	0	0	1.0	0	0

ing atmospheric nitrogen are listed in Table 25. Data concerning the fibrous roots and nodules of black locust are given in Table 26.

Mycorrhiza. When soil conditions are unfavourable, symbiotic root fungi may provide sources of nitrogen. Whereas in endotrophic mycorrhiza the fungus lives inside the cells, this is not the case with the ectotrophic variety (Table 27). Most authors hold that mycorrhiza further tree growth on poor soils. The fungi improve nutrition by providing additional mineral salts and especially by increasing available nitrogen, and they activate tree metabolism by supplying growth-promoting substances. They produce a 100—1000-fold enlargement of the root surface resorbing water and salts, but with an ample supply of minerals trees can exist just as well without mycorrhiza. For example, seedlings of *Pinus taeda* lose the mycorrhiza when the level of mineral nutrition is high (Brown, 1954). It is believed that mycorrhiza secrete antibiotics.

A tree's root system mostly lives in symbiosis with several species of fungi whose contributions to the system may change in the course of time (Lyr et al., 1967). Soils of the steppe and prairie contain no mycorrhiza, but this can be remedied by vaccination so that the trees may prosper. Afforestation of industrial waste heaps and dumps makes the same measure necessary. Hatch (1936, 1937) showed that pines grown in mycorrhiza-vaccinated prairie soils had absorbed 234% more phosphorus, 86% more nitrogen and 75% more

TABLE 27

Mycorrhiza on different kinds of trees (after several authors)

Ectotrophic types of mycorrhiza

Abies	*Betula*
Larix	*Carpinus*
Picea	*Fagus*
Pinus	*Populus*
Pseudotsuga	*Quercus*
	Salicaceae

Endotrophic types

Araucaria	*Acer*
Cryptomeria	*Alnus*
Cupressaceae	*Casuarina*
Ginkgo	*Liriodendron*
Metasequoia	*Platanus*
Sequoia	*Podocarpus*
Taxodiaceae	*Robinia*
	Ulmus
	Leguminosae

potassium than was absorbed by pines grown in mycorrhiza-free soils. The latter showed yellowish needles and poor growth. Dale et al. (1955) found that chlorotic *Pinus banksiana* trees were cured by a supply of forest humus containing mycorrhiza fungi. (Further literature in Zak, 1964; Harley, 1969; Marx, 1971.)

4.1.7. Mechanical factors of the environment

Wind. Moderate winds affect the life functions of trees in different ways, as already mentioned in the sections dealing with the individual economies. Mechanically, wind becomes effective when it reaches storm intensity or when it blows very frequently. White fir, larch, swiss pine, birch, mulberry, maple, oak, hornbeam and sycamore are considered storm-proof. To a smaller degree the same is true of spruce and pine, especially the white pine (*Pinus strobus*). Measures to prevent storm damage are discussed on pages 191 and 196. The use of cable winches to test the stability of trees must be rejected because this method causes serious root damage. Windfalls do not only occur in cultivated forests; they also play an important role in natural woods in rejuvenescence and in allowing an exchange of tree-successions (Stearns, 1949).

Plate 6. Snow-danger for trees (Photo Bernatzky).

TABLE 28

Heights and burdens of snow, measured during January 1941 near Zuerich/Switzerland (after Leibundgut, 1943)

Stock of trees	Snow depth on the ground	Snow burden t/ha on the stock
Open country	90	—
Spruce 15—20 years	33	855
Spruce 40—60 years	47	635
Spruce 70—98 years	48	630
Beech 15—20 years	57	495
Beech 40—60 years	59	465
Beech 70—90 years	64	390

Snow. This can damage trees in two different ways:
by snow blasting the sprouts peeping out of the snow,
by the weight of wet snow which may result in broken branches (Plate 6).
One mm^3 of rainwater corresponds with ca. 13 mm^3 of fresh snow. The weight would accordingly amount to 1 kg/m^2. The weight of wet snow may

Plate 7. Forest in Norway, near Hindsetter in Jotunheimen (Photo Kraus).

68

TABLE 29

Percentage of lightning of the different tree species (after Stahl and Moreillon in Walter, 1960)

Spruce	32 %
Fir	32 %
Oak	19 %
Poplar	15 %
Pine	14 %

be 3.3 times that of dry snow (Bonnemann and Röhrig, 1971). Leibundgut measured snow weights in Switzerland in 1943 (see Table 28). Tolerance of snow weight varies. Conifers with pliable branches are relatively resistant. Among deciduous trees those with thin twigs such as linden and birch are particularly vulnerable. Spruces and pines in areas approaching the alpine and arctic timber-line develop a more and more pointed, slender crown. What is apparently involved here — as in other unfavourable locations — is a retardation of the lateral shoots due to cold and drought (Plate 7).

Fire. Trees are often struck by Lightning (Table 29). Presumably trees with smooth bark (beeches) lead the rain water from the branches towards the trunk which then becomes moist. In this way the bark becomes conductive so that the lightning is led off down the trunk without damaging the tree. Trees with deeply fissured bark have a poorly moistened trunk surface. The lightning then penetrates into the water-saturated and conductive cambium and cleaves the trunk.

Fire is very dangerous to forests. Surface fires hardly ever damage trees with thick bark. Old trunks of *Pinus strobus* were destroyed only at cambium temperatures of 60°C lasting 2—4 minutes, and of 65°C for less than 2 minutes (Kayll, 1963). Trees with thin bark, however, are particularly vulnerable. Fires reaching tree crowns often destroy all vegetation.

4.1.8. Interference of ecological factors

The individual factors mentioned are always subject to marked alterations in intensity. Moreover it is not just one factor which determines optimum tree growth, but always a complex of different factors. Only in extreme locations can one single factor be decisive. Different factors may impede each other (e.g. root reduction caused by lack of light will intensify water stress in dry periods or on dry soils); but they may also compensate each other (a high nutrient level in the soil intensifies shade tolerance). A single factor such as heat or precipitation, is not decisive and does not enable us to predict the growth of a tree species or of an individual tree. For this reason

Walter and Lieth (1960—1967) and Walter et al. (1975) brought together climatic diagrams for a great number of places on earth to produce an atlas, on the basis of which a survey can be made of the most important elements in one single presentation.

4.2. Phytocoenoses

In nature plants and trees do not grow singly but always in mutual relations with other individuals of the same or of different species. Here, too, they depend on the ecological economies and biotopes, but further influ-

Plate 8. Natural plant community with *Pinus halepensis* (Israel) (Photo Bernatzky).

ences result from the fact that these groups of organisms are associated together as a unit (Plate 8). An essential characteristic of such a unit is the competition between species. The existence of individual tree species often depends on the presence or absence of competitors. Walter (1960) points out that e.g. *Pinus radiata* and *Cupressus macrocarpa*, both of which have an extremely small distribution area on the Californian coast near Monterey, showed the best growth characteristics when used for wind-screen planting and afforestation in Australia and New Zealand. In their new locality they became characteristic trees of the landscape. Likewise *Robinia pseudacacia*, a native of Eastern North America, developed into the dominant tree of the Ukrainian and Hungarian steppe and is looked upon as a characteristic tree of that landscape. The most likely reason is the absence of competing tree species in the new location. Evidently the specific genetics of a tree, that is its ecological constitution, is of importance for its distribution. Moreover, the distribution of plants on earth is also historically conditioned, so that geological and paleontological findings must also be considered.

All the plants of a territory form its vegetation. The study of vegetation includes plant sociology (or phytocoenology) which deals with plant communities as "environment-dependent combinations of plant individuals which are in reciprocal competition with each other and which in turn alter the environment" (Ellenberg, 1956). In this process, the single plant individuals maintain their independence; the plant communities are real communities. However, plant sociology in ordering the plant world works with abstract types and concepts. Its basic concept is that of association, defined as a community of a particular vegetative composition with uniform local conditions and a uniform physiognomy. It is characterized by its leading species. Differential species subdivide the associations into subassociations and variants. For their parts, the associations are joined in higher units such as formations, unions and classes.

Plant communities may be distinguished from each other on the basis of widely varying principles. Therefore no standard schemes exist for the arrangement of the plant world in terms of abstract plant communities. Plant communities may be differentiated on the basis of
physiognomic particularities,
floristic particularities,
the final stage of vegetational development (climax),
local peculiarities.
Accordingly a pattern of vegetation will be obtained by
— tabulated comparison,
— mathematical determination of relations of similarity,
— ecological grouping,
— areal-geographical pattern,
— dynamic-genetic classification (Ellenberg, 1956).
Thus a plant community may be defined as a type of plurality of popula-

tions, characterized by a particular combination of species and by certain standard local conditions. Research work on this subject has developed differently in different countries. For details see special publications (Braun-Blanquet, 1932, 1964; Ellenberg, 1956; Knapp, 1971; Wilmanns, 1973; Whittaker, 1973; Mueller-Dombois and Ellenberg, 1974).

Dynamic-genetic classification

Plant communities are not static but dynamic entities. Because they are the expression of all the local factors affecting them, they are bound to react to alterations therein. This takes place in stages. An example is the scheme, divided into stages, of the silting up of nutrient rich ponds in Northwest Germany (according to Tüxen and Preising, simplified). On an open pond a nymphaea community develops (first stage of the succession). This is followed up by various intermediate stages (rush and reed communities), and then by others dominated by Carex which, in the next stage, develop into willow and wild cherry scrub. The climax of this succession is a bog-wood dominated by black alder (progressive succession). Human interference, in the form of clearing which involves the removal of all trees and the scrub, causes regressive succession leading to a meadow (substitute community). If human influence on this meadow ceases (no pasturage or mowing) a secondary progressive succession will again turn the meadow into a bog-wood dominated by black alder. Successions therefore result from alterations, natural or man-induced, in environmental conditions. Moreover, the vegetation causes the location itself to undergo gradual change, which is favourable to some species and harmful to others. It is thus also possible to infer changes in soil condition from the condition of a plant community. Another succession scheme may look like this (simplified); waste land → rubbish heap → pioneer woods → pre-forest → mid-forest → end-forest (progressive succession).

In this connection the role of trees in the plant community is of interest. Let us take one example (Tüxen, 1937).

Order: *Fagetalia silvaticae* Pawlowski 1928; Union: *Fraxino—Carpinion* Tx. 1936; Association: *Querceto—Carpinetum medioeuropaeum* Tx. 1936; Subassociation: *Querceto—Carpinetum medioeuropaeum typicum* Tx. (1930) 1937. This subassociation comprises not only herbaceous plants but also the following trees: *Carpinus betulus* as the dominant species, as well as the species characteristic of the union *Prunus avium, Acer pseudoplatanus, Fraxinus excelsior, Prunus padus* and *Salix caprea*. The species characteristic of the order are *Fagus sylvatica, Rosa arvensis* and *Acer platanoides*. Escorts are *Quercus robur, Corylus avellana, Crataegus* species, *Acer campestre, Sorbus aucuparia, Populus tremula, Prunus spinosa, Quercus sessiliflora* and *Betula pendula*. The subassociation forms the climax on siliceous soils of the territory, especially on very fertile, dry loess. Almost everywhere, forests on this type of soil have been changed into fields for agricultural use on account of their high fertility.

The natural compositions of plant communities have also been changed by human land cultivation. The aim of modern plant sociology is therefore to determine the "potential natural vegetation" (sum of the permanent natural climax communities) for each territory in contrast to the actually existing vegetation which has been changed by man, i.e., the state of vegetation as it would develop by itself in various localities without human interference. Authors such as Wilmanns (1973) differentiate between the original vegetation which was there prior to human influence, and potential natural vegetation. The latter term denotes the community which would develop if human influence were to cease and if no other changes occurred. Then all ecologically effective factors would be in proper balance.

Despite all alterations in the vegetation pattern the plants that still exist (mostly herbaceous) make it possible to map a given area's potential natural vegetation. Its plant communities then determine the choice of the trees best suited to the territory concerned, as the above-mentioned example shows. Tüxen's (1956) concept of the "potential natural vegetation" is a clear and unambiguous basis for vegetation mapping which renders superfluous quite a few controversial conceptions such as climax, climax community and climax complex. The methods to be employed in preparing such vegetation maps are amply described by Tüxen (1956) and Trautmann (1966), and further references may be found there. For maps of world scope, areal cards are used (Meusel et al., 1965; Schmucker, 1942; Walter and Straka, 1970). Vegetation monographs have been written by Knapp (1965) for North and Central America, by Knapp (1973) for Africa, by Numata (1974) for Japan, by Metcalf (1972) for New Zealand, by Holliday and Hill (1969) for Australia, and by Sargent (1965) and Walter (1968, 1973) for North America. New literature is reviewed annually in Knapp, R.: Progress in Botany: Vegetation ecology.

The significance of plant sociology for practical planting
The riskier a locality is for planting and the less care that can be given to trees, the more important is a good knowledge of the area's potential natural vegetation. This is true regardless of what is to be planted. Forestry requires the proper use of locations, which in turn requires the proper selection of trees. Unsuitable species disturb the soil conditions. For instance, although douglas fir, red oak, white pine and poplars grow much faster than do species more suitable for a given locality, they do so at the expense of nutritive reserves in the soil. Rock decomposition cannot replace the absorbed nutrients in such a short time. Thus the soil will very soon be exhausted. A layer of matter that does not easily decompose will shut the air off from the soil. Acid humus will develop. This happens mainly with spruce and pine but also with beeches in all those localities where they would not have grown originally. Some trees change soil conditions completely by delivering substances that were formerly lacking. Locust, for instance, enriches the soil

with so much nitrogen that a change in the composition of the species results. Nitrogen-loving plants will be preferred and these will suppress the natural tree species or even displace them.

With the help of pollen analyses and forest-history research, plant sociology can determine which species are most suited to a certain locality. Certainly the cultivation of alien tree species may yield larger returns at first, but only at the expense of later damage with consequent production deficits which all too soon will use up the initial profit. Today, when afforestations of immense dimensions are needed and carried out all over the world, plant sociology is essential if failures are to be avoided.

Greening of wasteland. Trial plantings of trees on sterile soils mostly end in failure. Even hardy, simple types such as locust and white pine are not able to provide good soil conditions for the species of trees required later on. However, even on difficult territory a natural vegetation development proceeding in successions can lead to success. Scrub or shrubby communities, for instance, can bring about a forest with an abundance of trees in almost all localities in Central Europe. Pioneer woods in terms of their individual requirements include, in Central Europe, the following species: *Juniperus communis, Larix europaea, Pinus sylvestris, Pinus mugo, Populus nigra, Populus tremula, Salix* species, *Betula pendula, Betula pubescens, Amelanchier ovalis, Cotoneaster integerrima, Crataegus monogyna, Crataegus oxyacantha, Sorbus aucuparia, Rosa rubiginosa, Rosa canina, Sarothamnus scoparius, Acer pseudoplatanus, Hippophae rhamnoides, Sambucus nigra, Sambucus racemosa, Alnus incana, Alnus viridis, Alnus glutinosa* (Knapp, 1971).

Plantations in parks and gardens. Here plant sociology offers the same advantages as in the forest. Reduced costs and maintenance expenses for suitable trees are of additional importance. Danger of damage by drought or frost can mostly be excluded by the use of indigenous species which also fit far better into the aesthetic picture than bizarre forms or extremely bright colours.

Cultivated landscape. Everywhere man has changed the picture of the former natural landscape into cultivated land. Fields and meadows have taken the place of original forests and have added to the picture in many ways. Occasionally the unexciting uniformity of woods has been replaced by a delightful variety of views in a changing landscape, but investigations of cultivated land have shown that the ecological boundaries are still visible along the border between forest and field. For example, in Northwest Germany the original vegetation — that is, the tree stock that once existed — still influences the pattern of settlement and housing and that of farmland partition, use of the fields and also house construction. For an examination of the possibilities of using plant sociology for the insertion of streets and waterways into the landscape as well as for consolidation of dunes, see Tüxen (1961).

All this shows how problematic the usual catalogues of "Trees for every site" can be. They often consider the single features of the soil in isolation. Even if those lists may be a first aid for tree planting now and then (Hillier) they should not be used any more thoughtlessly than the inventories for different cities which are always in demand.

4.3. Trees in the landscape

4.3.1. Trees in forests

The forest is a perfect example of an ecosystem, though this is not to say that modern cultivated forests are undisturbed natural ecosystems. Nonetheless, the principal components of an ecosystem are to be found in a forest: the abiotic complex (radiation, heat and water economies, etc.) and

Plate 9. Natural forest (beeches, spruces, firs) in "Bavarian Forest" (Photo Kraus).

the biotic complex (producers, consumers, reducers). (Tschermak, 1950; Weck, 1956; Assmann, 1961; Lutz, 1963). The ecosystem is not completely closed. From outside it receives the sun's radiation, precipitation and minerals (dust). It disposes of water (ground and drain water) and releases energy through reflection, emission and convection. Furthermore, gas exchange takes place with the surroundings (O_2, CO_2, water vapour), but in a natural forest energy and matter are in equilibrium. The extent of man's exploitation of the forest is the extent to which this equilibrium will be disturbed (Plate 9). What has been said (pages 38—68) about the individual economies is equally true of the forest, except for some modifications concerning climatic conditions.

4.3.1.1. Radiation economy. Solar radiation impinges upon the crown surface of the forest, so that only a little of it can penetrate to the ground. Most of the radiation (80% on average) is held back on the crown surface and only about 5% reaches the forest floor. The following table shows the light intensity on the forest floor with different species of trees (Table 30). In tropical rainforests the light intensity on the ground can be reduced to 1% of the brightness in the surrounding open country. Part of this process is a filtration of the radiation, favouring the red zone. This zone of the spectrum is the most effective for photosynthesis. Reduction is more marked in the blue—violet zone. Thus the radiation balance is a different one on the forest floor, in the trunk area, in the crown area and on the crown surface.

4.3.1.2. Heat economy. Here again the classification just given applies. A specific temperature climate develops in the forest which differs from that of the surrounding open country by its well-balanced air-thermic conditions

TABLE 30

Light intensity on the forest ground in % of the light intensity of the surrounding country (after Geiger, 1961)

Tree species	Without foliage	With foliage	Authors
Beech	26 — 66	2 — 40	Lauscher and Schwabl, 1934; Naegeli, 1940; Trapp, 1938
Oak	43 — 96	3 — 35	Sauberer and Trapp, 1937, Geiger and Amman, 1931, 1932
Ash	39 — 80	8 — 60	Lauscher and Schwabl, 1934
Birch		20 — 30	Lauscher and Schwabl, 1934
Fir	2 — 20		Lauscher and Schwabl, 1934
Spruce	4 — 40		Lauscher and Schwabl, 1934; Naegeli, 1940
Pine	22 — 40		Lauscher and Schwabl, 1934; Sauberer and Trapp, 1937

and calmness. The roof of the crowns of a forest is the active surface, where the largest part of incoming radiation is absorbed and transformed into heat. There the highest temperatures are measured during the day-time. During the night, this is where emission and cooling off are most intense. With opening up of the forest, the thermic contrast with the open country diminishes. In the leafless state, however, these particularities are not applicable. Almost all the year round, but especially in summer, temperatures on the forest floor are lower than in the open country.

4.3.1.3. Carbon economy. This was discussed on page 45.

4.3.1.4. Water economy. This shows many peculiarities. In a forest only part of the precipitation reaches the ground. Another part remains on the leaves and evaporates there (interception). Only a very small part is directly absorbed by the leaves and the bark. In a coniferous forest the average interception loss amounts to 30%, in deciduous woods to 20% (Table 31). Dormant deciduous woods have less interception than do coniferous stocks. In summer the throughfall is about the same as with conifers, but the trunk downflow is distinctly larger in deciduous woods. In the leafless state in winter, deciduous trees have greater throughfall than do conifers. Table 32 shows details concerning transpiration of various plant stocks. Henrici (in Walter, 1960) investigated the water consumption of some trees and tree stocks in South Africa (Table 33). Planted tree species can be kept alive only if additional water is available. The consequence of this water balance is that even sparse stocks of trees can exist only if they receive a yearly precipitation supply of at least 110 mm, or 10—12 mm monthly (Larcher, 1973a). Tree planting and afforestation in arid climates are therefore very problem-

TABLE 31

Interception of coniferous and of deciduous woods (after Hoppe, 1896 in Bonnemann and Röhrig, 1971)

Quantity of rain (mm)	Interception during the summer in percent of open country precipitations		
	Spruce	Pine	Beech
5	71	48	38
5 — 10	57	38	24
10 — 15	44	23	19
15 — 20	31	25	13
20	24	8	10
Average of all measurements	41	24	20

TABLE 32

Transpiration of plant stocks according to measurements by numerous authors (after Larcher, 1973a, shortened)

Stock	Territory	Evaporation of the stock yearly (mm)	Precipitation yearly (mm)	Evaporation in % of precipitation
Forests and tree stocks				
Eucalyptus plantation	South Africa	1200	760	160
Tree plantation	Java	2300 — 3000	4200	55 — 72
Tree plantation	Brasilia	600	1400	43
Evergreen rain forest	Kenya	1570	1950	80
Bambus forest	Kenya	1150	2160	53
Mixed forests	Europe Japan, U.S.A.	500 — 860	1000 — 1600	50 — 54
Coniferous forest	Middle Europe	580	1250	46
	northern Taiga	290	525	55
Forest steppe	Russia	(110)200 — 400	400 — 500	(25)50 — 80
Macchia	Israel	500	650	77
Chaparral	California	400 — 500	500 — 600	80 — 83
For comparison				
Cornfields Meadows	Germany	about 400	800	50

atic. Similar observations (including a drop in ground water level) have been made in Eastern European steppes (Buchholz, 1950). Solitary trees can grow there, but a closed stock will be eliminated by the competition between the grass roots and those of the young trees. Wind-screen plantings can be successful in these locations only if they are protected from this root com-

TABLE 33

Water consumption and precipitation in Pretoria, South Africa (after Henrici, in Walter, 1960)

Kind of tree stock	Water consumption (mm)	Precipitation (mm)
Mixed stock without undergrowth	140	760
Pinus insignis, planted	760	760
Eucalyptus, planted	1200	760
Acacia mollissima, planted	2500	760

TABLE 34

Yearly transpiration of forest stocks according to several authors (after Bonnemann and Röhrig, 1971)

Tree species	Yearly transpiration (mm) according to		
	Polster	Pisek and Cartellieri	Sonn
Birch	564	388	
Beech	456	268	
Larch	564	379	437
Spruce	516	235	223
Pine	282	290	345

petition during their first years. Then *Robinia pseudacacia*, *Acer negundo* and different species of oak can grow successfully.

A forest consumes much water, as Table 34 shows. Investigations (which are not easy to carry out) have shown that forests consume far more water than does the open country or grass land. Therefore it is considered evident that a forest's water output is less than that of forest-free territory. This is not important in a humid climate but will become so in an arid one. By removal of the forest, the water output may be increased (Trimble et al., 1963; Keller, 1968, and references therein). On the other hand the forest has certain effects on the water. The stray and humus layer on the forest floor checks the surface drainage. Stray leaves work better than stray needles. Forest floors stay free of frost for a longer time in winter and can therefore absorb precipitation and thawing snow longer than can the open country. Thawing and water release continue longer. Moreover, the water flow out of the forest is more gradual and drainage fluctuations less marked, even in periods of drought. The water flowing out of a forest is of better quality and is cleaner, provided it has not been polluted by mineral salts or herbicides.

Measurements in forests have shown that air humidity within the stock is some degrees higher than in the open country. The denser the forest, the greater the difference, especially on high radiation days. Evaporation diminishes with increasing stock density as well as towards the ground. This increase of relative air humidity is largely due to the lowering of air temperatures and, to a smaller degree, to the increase in vapour pressure through evapotranspiration.

4.3.1.5. Mineral-substance economy. Generally forests grow on less valuable soils. This is especially the case in Europe and North America, where better soils are normally utilized for fields. A high turnover of minerals is typical of a forest. According to Fiedler and Hoehne (1965) the yearly quantity of absorbed nutrients per ha amounts to 25—45 kg N, 3—6 kg P, 11—33 kg K and 20–40 kg Ca. Ovington (1962) found even larger quantities in a 55-year old forest of *Pinus sylvestris*. Compared with trees in city streets, forest trees are far better supplied with mineral nutrients, if only because of the leaves or needles that fall to the ground and improve the soil.

4.3.1.6. Interference. In a forest the reciprocal influences of the individual members of the biocenosis are especially intensive. So far as plants are concerned, there is competition in the struggle for light, and for roots there is competition for water, nutrients and air in the soil; there is also competition for the effects of some metabolic products of the plants on each other. So far as animals are concerned, there is the damage done by noxious animals and pests, and the shelter offered them by the forest vegetation. Because the forests of the highly civilized world are artificial products and serve the production of timber, ecological damage is inevitable. Pelisek (1974) deals with the consequences of monocultures in forests. Muench (1972) and Schuett (1974) deal with the effects of the use of herbicides. Lutz (1963) considers all of these questions. The growing number of city dwellers seeking recreation in the forest leads to a particular ecological problem which consists in the soil's becoming more compact and subsequent impairment of the flora (Liddle, 1975).

4.3.1.7. Protective effects of the forest. Nowadays the importance of forests for the environment is estimated more highly than their economic value for timber production. Abundant literature is available on this subject (Turner, 1968; Keller, 1971a; Fontaine, 1972; Hasel, 1971; Bichlmaier and Gundermann, 1974; Leibundgut, 1975 to name only a few). Forests are eminently suited to serve as protection against erosion by water and wind. Ample proof of this can be found all over the world, wherever forests are cut down. In the mountains, forests are a good protection against avalanches. Their wind-screening effects are discussed on page 141.

In particular the forest plays an eminent role in the protection against immissions. Undisputed is its filter effect against particulate pollution (page 135). In comparison with solitary trees, tree groups or protective plantings, forests modify the filter effects. Incoming polluted air is lifted up in front of a closed forest and led away over it. A filtering then occurs in the turbulences which develop above the crown surface. This filtering effect is permanent, but it becomes relevant for humans only when they live in the immediate vicinity of the forest. Forests which are far removed from residential areas have no direct effect.

The term 'Immissionsschutzwald'' (immission-protective forest) was coined in West Germany. Two zones are distinguished. Zone I includes all forests which are themselves damaged by toxic pollution (e.g. Brandt and Rhoades, 1973) but which are nonetheless of particular importance for immission protection. Zone II includes those forests which reduce the concentra-

TABLE 35

Protective distances between industrial plants and forests or residential areas in North Rhine-Westphalia (shortened)

Kind of industry	Protective distance
Brickworks, pottery works, concrete and morter factories, wood processing, spinning mills, etc.	300 m
Cement and gypsum industry, lime kilns, rolling mills, metal works, pressing rolls, steel industry, slaughter houses, etc.	500 m
Foundires, intensive animal breeding, engine works, etc.	800 m
Hammer mills, refuse incinerators	1000 m
Oil refineries	1200 m
Blast furnaces, steel industry	1500 m
Chemical industry, petrochemistry, power plants, etc.	2000 m

tions of noxious substances without suffering any real damage themselves (Wentzel, 1963; Forstl. Bundesver. anstalt Wien, 1971, 1972; Knabe, 1971, 1972, 1973a, b). Afforestation of new immission-protective forests is recommended at certain distances from industrial areas (Table 35). Where the factory or plant is at a smaller distance than the minimum shown, the forest is associated with zone I, otherwise with zone II.

The trunk area of a forest is almost free of aerosol and gas. According to Burger, 1953 (in Keller, 1971a) a stock of beeches can filter 280 kg/ha of dust, stocks of oak 540 kg/ha, and stocks of spruce 420 kg/ha. According to Steubing and Klee (1970) the mountain pine even filters up to 1000 kg/ha. For correlations between forest and gaseous pollution see page 140.

Carbon dioxide consumption by the forest. Land-clearance and combustion processes are increasing and this causes a rise in the CO_2 content of the air. "A substantial increase in forest areas and cubic volume of wood in the forest would reduce the carbon dioxide content. About 2.7 billion acres of new growing forest (120 percent of the 50 states of the U.S.A.) would be required to absorb the carbon dioxide generated by fossil-fuel combustion at the rate prevailing in 1962" (Peterson, 1970b); see also Lange and Schulze (1971).

Noise protection. A multi-staged, dense young forest reduces the noise per meter by 0.16 dB; thus a forest of 200 m depth can achieve a reduction of

32 dB, and one of 250 m depth a reduction of 40 dB. This is also the minimum level for the protection of residential areas. Dispersion of the noise is influenced by meteorological conditions. The dispersion of sound is increased in territories with temperature inversions (valleys, troughs). Noise produced in the forest (power saw, traffic noise) is felt to be disturbing at far greater distances (Wendorff, 1974).

4.3.1.8. Recreation forest. The forest is one of the commonest and most popular recreational areas (Brockmann, 1959; Barthelmess, 1972; Bernatzky, 1973c). However, it is not easy to quantify its recreational functions because we are here dealing with biological processes. Life and what sustains it usually eludes easy comprehension, because the sum of all the details measured by no means constitutes the whole that we call life. This shows particularly clearly that material, mechanical modes of comprehension are inadequate. We shall be more successful if we point out the overall effects produced by forests (Flemming, 1972b).

Good air. The forest produces large quantities of oxygen (page 155 onwards). It is true that it consumes almost the same quantity for the decomposition of dead matter, but forest air is still exceptionally rich in oxygen. Forest air contains large quantities of volatile oils (terpenes). These aromatic substances favourably affect the forest visitors. They deepen respiratory intensity. Amelung (1952) assumes that the forest air contains particular admixtures, which he calls "air-vitamins" and which have a favourable effect on the human organism. Forest air is generally free of noxious gases since there are no factories in the forest. Forest air contains less dust than does the air of the open country and city (page 135 onwards).

Radiation. The radiation of the sun has thermal, chemical and psychical effects on humans. In the forest these are present in a very favourable mixture because of the alternation of glades and lower and higher stocks of trees. That is why it would be wrong to afforest all glades.

Thermal effect. This is composed of various climatic elements: radiation, wind, air-humidity and air temperature. Under the forest's influence they jointly constitute a gentle climate of weak stimulation. This is particularly beneficial to persons recovering from illness.

Acoustic effect. Normally the forest is free of noise. The chief sounds are natural ones such as bird song, the murmuring brook, the rustling of leaves and branches — all of them sounds which have a calming effect. The environment of the forest is the natural landscape which favourably affects the heat economy of the human organism, its vegetative regulatory mechanisms and its cardiac and circulatory activity.

If the forest is to become an established recreational area it should be accessible from residential quarters. Accordingly we may distinguish between the recreation forest in the immediate vicinity of the city, to be reached in half an hour or even less, the week-end forest at a distance of about 30—50 km, and the holiday forest. The recreation forest should receive proper care. Impenetrable thickets are not suitable for it. Marked walls, benches and refuge huts should make it accessible. Opinions vary on the composition of the tree species: some favour a light, deciduous forest, others a dark, severe coniferous forest. However, it is not these preconceived opinions which should determine the choice of tree species, but consideration of the ecological principles governing the forest's particular situation. If this is carefully considered the forest will itself provide the necessary and possible variation in its growth and form. It will be generally acknowledged that large clearings and telephone-pole forests can never serve as recreational surroundings.

Plate 10. Very old *Acer pseudoplatanus* on its natural stand near Tegernsee/Upper Bavaria (Photo Kraus).

Recreation centres in the forest. The average city dweller of today has lost his ties and relations with nature, and thus also with the forest, to a high degree. He looks for something like urban life even in his recreation, but here lies a great temptation. Of course the recreational forest should be attractive to its visitors. Opportunities for sporting activities should be there: minigolf, table tennis, boccia, open air board games, bowling, barbecues and other attractions; but these facilities should not be dispersed all over the forest. The following proposal presents itself as necessary mainly for holiday and national parks. The whole recreational area is divided up. First is the arrival area with parking space and opportunities for refreshments and for sport. Next, there is the strolling and picnic area, and finally the area for resting, where it is perfectly quiet (Plate 10). This last area should be left as natural as possible and be reserved above all for people who, preferring the undamaged ecosystem, want to benefit without disturbance from the forest's most important recreational factor that is, to stroll and rest in clean air and in a setting that stimulates the mind. Nor should the importance be forgotten of offering new impressions to visitors, by the clearing of panoramic views from hill tops or mountain peaks, riverside banks or slopes on lakes in order to make their wanderings even more beneficial to them. Noisy recreational facilities do not belong in the forest. They should not go beyond its edge.

The increasing use of forests for recreational purposes creates a particular ecological disturbance in the form of damage done to certain plant communities and single species as a result of trampling. Liddle (1975) classifies the results of investigations already carried out on this subject and records them via several measurable parameters.

4.3.2. Trees in parks

Parks in the landscape should be distinguished from urban parks or parks in urban areas with houses and buildings (Plates 11 and 12). The landscape park has similar ecological conditions to a light forest. The presence of larger lawn areas surrounding the trees already makes the forest climate more like that of the open country, with higher temperatures and lower relative air humidity. The more lawns there are in the park the less favourable will the soil conditions be for the tree roots (page 28); poor soil aeration and the removal of fallen leaves result in an interruption of the mineral cycle. Parks in urban areas are subject to the disadvantages of the urban climate, that is, to higher temperatures, lower air humidity and reduced moisture in the soil. They are also harmed by the pollution of urban air and by such industries as may be nearby. The extent to which urban parks can ward off these harmful effects or even reverse them, thus improving the urban climate, depends on their dimensions, their quality and maintenance.

The larger the parks, the better off they are. Minimum dimensions yielding clear and observable effects on the climate lie between 1/2 and 1 hectar (Sperber, 1974). The most essential part of parks and green areas will always

84

Plate 11. Recreation under park trees (Frankfurt/Main, W. Germany) (Photo Bernatzky).

be the trees. The better the ecological requirements are met, the more will the beneficial effects of the trees be felt in the town's climate, but where lawn areas grow and tree areas diminish, the beneficial effects of trees diminish as well (Bernatzky, 1960; Sperber, 1974; Brahe, 1975). The ecological benefit of parks depends on their layout in the whole topography of the town. It makes a difference whether they are situated in moist valleys close to ground-water level, or on arid slopes with poor soil and little water. These differences must be observed in selecting the species of trees, and here again the potential natural vegetation gives the best advice. The use of exotic tree

Plate 12. Park in the landscape. Villa d'Este in Tivoli, Italy (Photo Bernatzky).

species requires the greatest care (page 128). The best way to plant them is in ecologically related groups and not one by one, dispersed all over the park. Parks and green areas with trees are among the basic requirements of towns. Their significance is manifold:
they serve a human need,
they improve the climate,
they may turn into the leading idea of planning (garden city),
they are a formative element of urbanity,
they serve the display of representation,
they symbolize preventive care for health,
they are part of tradition,
they belong to public utilities along with sports grounds, cemeteries, children's playgrounds and recreation centres (Klaffke, 1972).
The city signifies for its inhabitants their detachment from the country and from nature. Today's cities have grown so much that the open country can only be reached by a great loss of time, so that people rarely go. Possibilities of direct contact with nature are hard to find. Open mindedness towards nature is a thing of the past. Trees and green areas are effective in counteracting these developments. They stimulate the aesthetic sense and so produce a beneficial effect on mental and physical health.

Groups of trees and tree-covered areas can make a decisive contribution to the improvement of urban climate by restoring natural conditions and original temperature values as much as possible. They are the only possible means of reducing overheating (page 145). They are supremely important for restoring the ecological balance. They should therefore be situated where their specific effects are most needed and that is especially in urban centres (Bernatzky, 1972). After all, on the outskirts of a town nature and a natural climate still exist. Green areas are not interchangeable. This is to say that trees and tree-covered sites must be situated where they are needed climatically and not just be degraded to cosmetic arrangements. If buildings are massed around small green areas their effects will soon be lost (Sperber, 1974). The smaller the number of trees on a green area the lower its effective value. This is particularly true of small ornamental squares with a little lawn and undersized decorative shrubby. In Central Europe and East Asia parks often have a historical and cultural value. They keep alive the memory of famous persons in philosophy and religion (East Asia) or of the fortified walls encircling medieval European towns which in modern times have been transformed into green parks (Bernatzky, 1960). Many an old park was once a private palace garden.

Green areas should never simply be a use for otherwise useless leftover strips resulting from street planning. Apart from the fact that ecological conditions there are very unfavourable, green plantings in these locations lead to a devaluation in the idealistic sense. From the beginning, their planning must be an integral part of town planning. On leftover strips, however, single trees may be planted after thorough improvement of the site. Incorporation of park planning into town planning has happened in the garden cities, but elsewhere consistent park planning stagnates very much because of the hitherto solely aesthetic evaluation. Related disciplines like geography and meteorology must participate. Fundamental scientific papers already exist (Bernatzky, 1972; Greiner and Gelbrich, 1974; Sperber, 1974), but their practical realization in towns leaves much to be desired.

4.3.3. Trees on traffic routes (Plates 13—16)

Trees have always lined streets. Remember Napoleon who planted trees along all his supply routes across Europe and right into Russia to make them recognizable. Today's heavy traffic, however, gives rise to the objection that trees cause numerous accidents and casualties. Certainly traffic is necessary. The motorized vehicle is an essential element of freedom and the street is there primarily for transport. Yet it is not the street trees which endanger traffic, but the drivers themselves. Where there are no trees at all drivers crash into buildings, bridge rails, poles carrying overhead wires, and other cars. Well-documented investigations available today show that trees do not necessarily cause accidents. Thedic (1959) investigated the RN 83 between Bourg and Lyon (France). He found the accident rate highest where trees

Plate 13. Trees show the street direction in the best way (Photo Bernatzky).

Plate 14. Trees give the landscape a special value. Morvan in Burgundy/France (Photo Bernatzky).

88

Plate 15. Trees are landmarks in winter (Odenwald, Germany) (Photo Zinsel).

were right on the edge of the road. However, where the distance between the edge of the road and the trees was more than one meter, a lower rate than on treeless streets was found. These findings were confirmed by Bitzl (1968). He points out that on numerous occasions trees have prevented accidents or checked their consequences by keeping vehicles from crashing down a slope, protecting pedestrians from runaway cars, etc., but no statistics exist concerning these accidents which did not occur.

Some very interesting investigations were carried out at the Institute for Psychology of the University of Aachen (W. Germany), and reported by Boeminghaus (1974). His point of departure is that on technically perfect roadways one has to expect a rise in traffic accidents. The reason is, according to Boeminghaus, that while careful consideration is given to the technical details of road construction there is nothing but deplorable neglect of the driver's mental condition, especially his or her perception. The driver's capacity to receive optical stimuli is limited. He can only employ a small part of that capacity for the borders of his driving lane. On the other hand a continuous minimum stimulation is necessary since otherwise boredom and

Plate 16. Trees are landmarks in summer (Morvan, France) (Photo Bernatzky).

fatigue set in. A sudden obstacle or a complicated traffic situation then triggers off an impulsive reaction which causes an accident. The objects along the street must be familiar to the driver so that he can react to them without too much thought. The more easily this can be done, the safer the ride will be. If the objects along the road are completely new, the driver's attention is diverted from the actual driving activity. This reduces driving safety. A very familiar biological reference point is the tree along the border of the road. Since earliest times, trees have been used as symbols and signs of orientation. Moreover every single human being knows the appearance, size and form of trees since his or her early childhood. Trees can be variously arranged in ever new constellations of form and colour so that they never produce boredom and weariness. The succession of light and shade changes them continuously into always new yet still familiar figures. In winter, with their dark filigree of branches in the silvery white landscape, they often form the only point of reference. Thus from the psychological point of view we shall never grow accustomed to the multitude of biological stimuli. This is particularly noteworthy in view of the bleakness, emptiness and boredom of modern constructions employing artificial materials, for instance city roads and highways. In addition, the aesthetic effect of the living tree has to be considered in contrast with the often unsightly dead-looking building material in our mechanized environment.

Trees can indicate important changes of climate, such as sudden gusts of wind, before the driver arrives at the place, so he can prepare himself. The most important contribution of trees towards improvement of safety on the roads is that they serve as optical guides along the driving lane. Repeatedly we come across sections of roads which are not clearly recognizable even with traffic signs. In treeless streets the driver confronts them entirely unprepared. The effect of trees following the line of the road is to help to make driving less risky by facilitating early recognition and estimation of the traffic picture by the quick detection of directly approaching vehicles and a more exact measurement of distances and of the speed of the driver's own car as well as that of the approaching vehicle. The investigations cited leave no doubt about these effects. Trees were found very useful in alerting drivers to the diminution of their visual field with increasing speed (Fig. 28). The results may be summarized as follows:

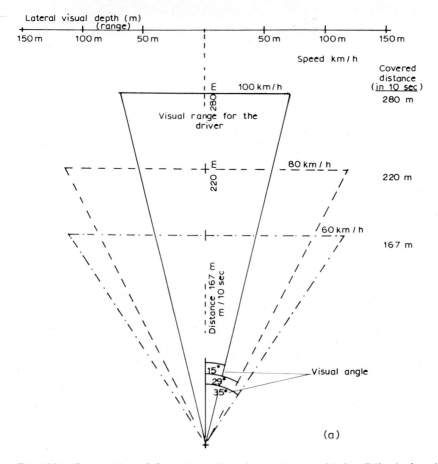

Fig. 28(a). Contraction of the optic angle at increasing speed (after Böhminghaus).

(1) The introduction of the tree as a biological factor brings about an alteration of perception;

(2) In tests, plans of streets with trees and shrubs were definitely preferred to those without vegetation;

(3) Streets with trees are recognized sooner than those without them;

(4) In tree-lined streets vehicles and their movements can be recognized sooner and better;

(5) Street directions are observed more accurately;

(6) In the visual field of the street, distance and speed can be measured far more accurately where trees serve as reference points.

The harder an area is to survey, (e.g. in the mountains) the more important for traffic safety are the trees. In such surroundings treeless streets often appear harshly destructive of the scenery, particularly when cut into a slope. The maximum gradient for rural roads is now generally 4% (3% for motorways). Therefore extensive work such as cuttings and embankment slopes is often necessary to maintain the roads at acceptable gradients. Planting indigenous trees and shrubs seems to be the best and most economic way to stabilize and camouflage them.

Seasonal changes always give new impressions of trees and landscape and thereby raise the effective value of roads in the landscape (Lorenz, 1952).

(b)

Fig. 28(b). Optical improvement of traffic conduct by trees and shrubs.

92

On pages 141—143 an explanation is given as to how trees and shrubs can protect the roads from wind and snow (see also Ortner, 1964).

For all these reasons the question should no longer be whether to plant trees along roads or not, but at what distance from the borderline they should be planted to avoid accidents and to promote road safety. The investigations and publications cited have shown that the minimum distance on general roads should be two meters (Fig. 29). For high-speed motorways a distance of 4.50 m is considered sufficient in Germany (by "Deutscher Rat fuer Landespflege", 1968). If the distances chosen are too large the driver cannot recognize them clearly because of his narrowing visual field. Dunball (1972) points out that it is sometimes argued that rows of trees shut off the road from the surrounding landscape and that only after felling the trees will

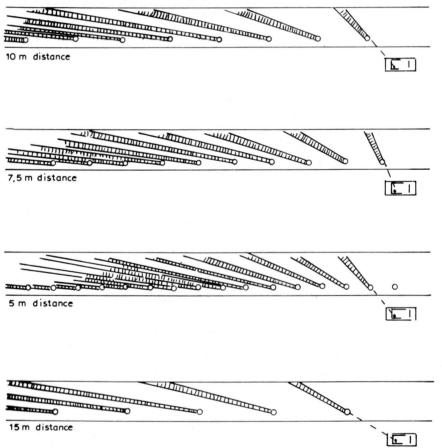

Fig. 29. The effect of tree spacing on the car driver. Narrow spaces act as a non-transparent wall; minimum space between trees 10 m.

the road become part of the landscape. Only in the desert do trees not belong to the landscape, but there is precisely where they would become the best traffic signs if it were ecologically feasible.

Many difficulties stand in the way of planting trees along roads and motorways. Infertility of the soil and its compaction, the bleak and exposed position of many of the sites, the effect of salt on plants and poor growing conditions (Dunball, 1968). However, success has been achieved with indigenous species in terms of quick root-taking, growing on, after care and aesthetic effect. Physical rather than chemical conditions are the limiting factors at the planting stage. Therefore, work with the planting drill is very problematic because a thorough soil preparation is not possible. Since motorways are massive structures over 30 m (= 100 ft) wide, plantations must also be massive if they are to be in proportion. Specific difficulties arise when the centre strip on highways has to be planted because in most cases it is found to be far too narrow. The minimum width of such a strip should always be 4 m. Only indigenous shrubs should be used, not trees, and only salt-tolerant species should be chosen (Kurusa and Brause, 1974).

4.4. Trees in the town

4.4.1. The ecological situation (Plates 17—20)

Light factor. The light received by trees in urban streets varies from almost unlimited to almost zero. Air pollution greatly reduces radiation (page 130). Furthermore buildings vary the distribution of the sun's radiation by their dimensions, their orientation and the width of the street (Fig. 30). Two sides of the same street may differ greatly. Thus the north side of a street running east—west receives far more light than its south side. In streets running north—south, light and shade values are almost equal. Each intermediate direction has corresponding values. Unequal exposure leads to one-sided crown development. A limited distance between a tree and a building forces it to grow asymmetrically into the street (Fig. 31). However, diminution of radiation is partly compensated for by the reflection from street pavements and buildings (for reflex numbers see Table 4, page 37). Tree species react differently to the diminution of light depending on their light demands (Table 5, page 38) and their shade tolerance (Table 6, page 39), but lack of light always leads to root reduction (Table 7, page 40).

Heat factor. Towns are about 0.5—1.5°C warmer than the surrounding country (yearly average). The heating up of street pavements and buildings, which form an artificial watertight rock, far surpasses the temperatures of plant-covered areas (Fig. 32). As a result of the higher temperatures, Spring starts earlier and Autumn ends later; the frost-free period is extended and overall temperature extremes moderated (for examples see Kratzer, 1956). These factors, which favour the growth of trees as well as, sometimes, the

94

Plates 17 and 18. City streets (Frankfurt/Main) — no comment needed (Photos Bernatzky).

Plate 19. Pines in Kyoto, Japan, Katzura-River (Photo Werkmeister).

selection of tree species, are balanced by the reduction of humidity resulting from rising temperatures. In summer it can get too hot for the trees as a result of irradiation and reflection, limited water supply, and the calm and therefore limited transpiration which reduces the heat-exchange of the trees. This is very often followed by scorching of the leaves and the bark at the foot of the trunk. Temperatures can get very high, especially in front of south-facing walls (Fig. 33).

Various measures have been tried to protect the foot of the tree trunk against heat from below. The best method proved to be painting with white latex (Lagerstedt, 1971). White protective collars made of cardboard or a similar material intensify the heat because of reflection. Tests with white latex showed that with an ambient temperature of $90°$ F the cambial area temperature of the painted stem was $108°$ F and that of the unpainted stem was $128°$ F. This means a temperature reduction of $20°$ F.

Heat-sensitive species are many lindens, horse chestnut, sycamore maple, Norway maple and hornbeam. Heat-resistant species are hawthorn, poplars, oaks and locusts. Between these groups are birches and plane trees. Trees with hairy leaves reflect a larger amount of radiation (ca. 30% as against 10%) so that their heating is reduced.

Water budget. On account of the sealing of street and roadway surfaces, the water from precipitation is drained off almost completely into the sewers and therefore is not evaporated by plants. The trees do not receive enough water, and the water balance is negative. Higher temperatures, less effective precipitation and reduced evaporation lead to a diminution of relative and

Plate 20. 300-year-old linden tree in Frankfurt/Main (Photo Bernatzky).

absolute humidity as compared with the open country. The area of open soil available to the trees which is not covered by street pavement is so small that the incoming precipitation there is not able to supply the trees sufficiently for their growth. Ground water is hardly ever accessible to the roots of urban trees, mostly because of construction work reaching far underground such as garages, railways and other subterranean installations. Theoretically, roots follow the water and keep on growing as the ground water level drops

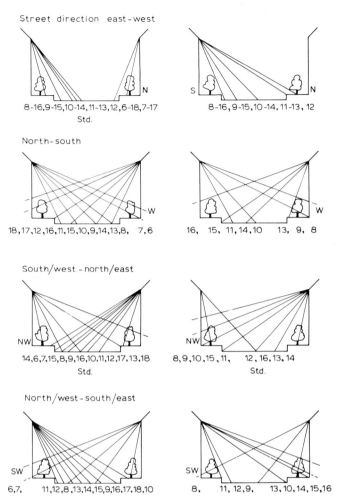

Fig. 30. Shading effect during one day with different street directions. The numbers give the hour of day (after Hoffmann, 1954).

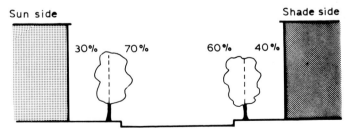

Fig. 31. Deformation of a tree (linden) resulting from one-sided exposure (after Grochals-kaja, 1952; in Hoffmann, 1954).

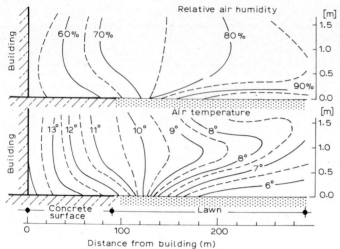

Fig. 32. Different temperatures above concrete surface and lawn in front of buildings (after Knochenhauer, 1934).

slowly, but they can never cope with the rapid drop that occurs with underground construction work. Thus we now may assume that ground water is no longer available to street trees and that the capillary rise of water in the ground is cut off.

Where will the trees get the necessary water? We are left to assume that the condensed water (page 52) is enough for reduced tree growth. Thus artificial irrigation becomes a necessity, since the water balance of trees must

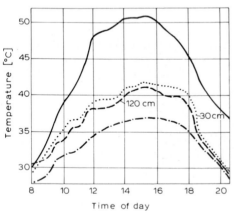

Fig. 33. Temperatures above an asphalt street (after Hoffmann, 1954). ———, the surface; ······, at a height of 30 cm; — — — —, at a height of 120 cm; — · — · —, by the side of the street.

be supported. After all, the beneficial effects of trees on humans are impossible without ample water supply. Meyer (1972) mentions an example of the results of reduced water supply on a tree of *Fagus sylvatica* 'Atropunicea', where the specimen standing in the pavement of a pedestrian walking area had smaller leaves and a larger number of stomata per mm², that is 515 per mm² on the south side of the crown and 382 per mm² on its north side. For comparison, a tree in the park had larger leaves and smaller numbers of stomata; 322 per mm² were counted on the south side, and 240 per mm² on the north side.

Given the precarious water balance of urban trees, only those species should be chosen which tolerate periodical water deficits in their natural habitat. Usually these are species with deep, extensive roots and with firm, glossy or hairy leaves (improved reflection of radiation and reduced transpiration). This group includes *Robinia pseudacacia, Ailanthus altissima, Platanus orientalis, Platanus hybrida, Quercus rubra, Quercus macrocarpa, Quercus robur, Corylus colurna, Celtis occidentalis, Ginkgo biloba, Tilia* × *euchlora, Tilia tomentosa* and *Sophora japonica*.

Plate 21. Street asphalt causing the death of a tree (Photo Bernatzky).

Carbon budget. The strained water balance considerably diminishes the production of matter and with it the growth of urban trees.

Soil. Very rarely do street trees stand in natural soil. Mostly they have to put up with sterile substrata like cinders, ashes or ruins; raw soils lacking humus and nutritive substances. Also, removal of fallen leaves in the Autumn blocks the circulation of mineral nutrients. As a result of the high lime content in the soil of old towns the pH-value is frequently above 7. For most trees, however, and most mycorrhiza fungi, it should be below that. For this reason the following pioneer woods are suitable for planting in such poor soils: *Robinia pseudacacia, Sorbus intermedia, Crataegus orientalis, Betula pendula, Populus tremula, Alnus incana.* The sealing of the soil by concrete,

Plate 22. Beech killed by road and park area construction near Aachen (Photo Brahe).

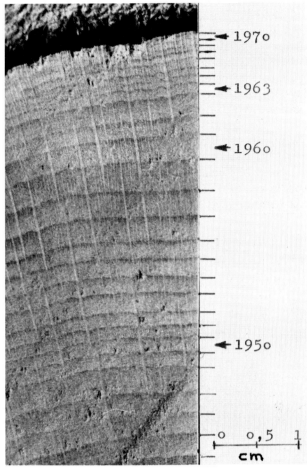

Plate 23. Annual rings of the felt beech. Road and park area were built in 1962. The annual rings are smaller than normal since 1963 (Photo Brahe).

asphalt, etc. reduces the gaseous exchange between atmosphere and soil. The compact surface has a diffusion resistance harmful to the oxygen supply of the soil. Oil dripping out of parked cars also causes airtightness of the soil (Plates 21—23).

Mechanical factors. Great damage is done to trees by construction work, impermeable street pavings, overfilling of the root areas with rubble and earth for the purpose of leveling, and the placing of supply lines near trees. Measures against this damage are listed on page 217 onwards. Overhead wires destroy tree crowns. Additional damage is caused by traffic. The root collar area can be injured by trucks, and injuries to bark and branches can be

102

inflicted by colliding vehicles. These injuries are dangerous because their seriousness is not recognized (cutting-off of conducting vascular bundles) and hence is neglected. Measures against this damage are listed on page 184 onwards. Especially harmful to the thin bark of young trees is dog urine. Installation of protective baskets has been tried; they are helpful but unsightly and expensive. Lately painting with lacbalsam (page 187) has proved successful against dog urination. Wind fells root-damaged trees in streets more easily than healthy ones; it also breaks large branches, especially of species susceptible to wind. Bracing and cabling may help (pages 191 and 196).

Investigations on dug-out trees. In Berlin, Hoffmann (1954) investigated the root structure of two specimens of *Tilia europaea* which had been carefully dug out. The trees were standing at a distance of 75 cm and 67 cm,

Plates 24—27. Excavation of trees in Berlin. 24, Tree (linden) on the border line. 25, The dug-out tree roots. Curbstone on the left. The roots are going to the right side (= under the sidewalk). 26, Roots turning away from the curbstone in parallel direction. 27, Roots turning away from the curbstone (Photos Hoffmann).

respectively, from the curbstone on the sidewalk. Except for very small spaces of less than 1 m², the root areas of both trees were totally covered with small paving stones and slabs. After having been dug out to a depth of 1 m, neither tree showed any roots growing towards the area under the asphalt of the street adjoining the curb. In both cases the roots clearly turned away from the curbstone and grew on in a direction parallel to the street. The possibility of damage to the roots during the operation of laying the curbstones was discarded after careful investigation. Instead everything pointed to the fact that the factor limiting root growth here was reduction of gas exchange. It was certain that the trouble did not stem from the water conditions as the soil underneath the asphalt was moister than the remaining root area. Only at greater depths (1.50 m), did some individual roots turn toward the asphalt. No harmful effects from the small paving stones and the slabs covering the root area were observed (Plates 24—27).

Chemical air pollution. We know that trees and other plants suffer from the effects of dust and gases in the air (Table 36). When the substances mentioned settle on the particles (kernels) of air as they do, they become attached to the leaves of trees or are deposited there by slight rain or fog. Under moist conditions the chemical substances become solutions which destroy the leaf tissue. The sturdier and glossier the leaves, the easier the dust gets washed off by heavy rain. Soft leaves, or those with hairs on the upper side, are always exposed to danger. In general, dusts injure trees less than do toxic gases. Dusts can exclude light from the leaves, corrode them, poison the soil, and make the air muddy. The full extent of the damage caused by these factors is not yet known (Garber, 1967, 1973; Brandt and Rhoades, 1972; Dunger, 1972. See also Håbjørg, 1973a,b; Bovay, 1972; Mellanby, 1972; Lowry, 1972; Smith, 1974). The phytotoxic gases get into the inner leaf via the stomata where they impair gas exchange by impeding photosynthesis and destroying the chloroplasts. High concentrations of SO_2 (1 p.p.m.) and HF (0.1 p.p.m.) cause whole leaf areas to die, but the chronic presence of lower concentrations (0.05—0.2 p.p.m., Larcher, 1973a) will also

TABLE 36

Heterogeneous matter in the free atmosphere

Particulate matter	ashes, ZnO, $PbCl_2$
Sulphur combinations	SO_2, SO_3, H_2S
Nitrogen combinations	NO, NO_2, NO_3
Oxygen combinations	O_3, CO, CO_2
Halogen combinations	HF, HCl
Organic combinations	aldehydes, tars, hydrocarbons
Radioactive matter	radioactive gases, aerosols

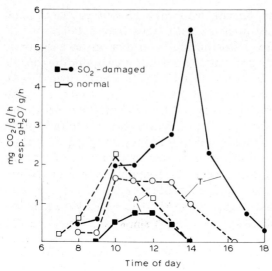

Fig. 34. Daily course of assimilation (A) and transpiration (T) of SO_2-damaged and healthy spruce branches (after Koch, 1957, in Keller, 1971b). The damaged branches show decreased assimilation; however, from 10 a.m. onwards there is highly increased transpiration (water loss!).

cause serious damage. Sulphur dioxide paralyses the movement of the guard cells and so causes excessive transpiration (Fig. 34). The tree's entire water economy breaks down this way and, in the extreme case, leaves and branches dry up. Permanent amounts of lower concentrations cause reduced increment (Wentzel, 1971). As a result the whole growth habit of the damaged trees changes. The foliage (or needles) gets thinner, growth lessens, crowns and trunk structures are distorted (Fig. 35). Vertical growth is more severely affected than the increase in thickness. Pines and deciduous trees flatten their crowns, thus keeping out of the high strata of particularly heavy pollution, but trees with monopodial growth, such as spruce and fir, are not able to do this. On comparable soils, air pollution reduced the potential growth of pines at the age of 70 years to a maximum of about 1/3 of the height, 1/2 of the medium trunk diameter, 1/10 of the wood mass and 1/14 of the wood value. Growth inhibition begins when the trees are about 20 years old and 5 m high. (Compare also Kozlowski and Keller, 1966; Dochinger, 1971; Greszta and Olszowski, 1972; Vins and Mrkva, 1972.) Fluorine is about 100 times more poisonous than sulphur dioxide so that very low concentrations are enough to cause damage: leaf edge necroses, ship-like turning up of the leaf blade, dried up needle points on conifers. All this amounts to a drastic reduction of assimilation organs (Keller, 1971b, 1973c,d, 1975a,b; Bohne, 1964, 1972; Berge, 1976; Wentzel, 1965).

A special form of immissions are the oxidizing kinds of air pollution, caused mainly by the photochemical reactions of exhaust-gases from motor

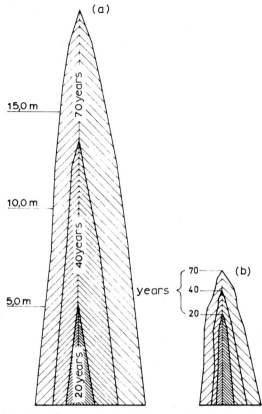

Fig. 35. Trunk analysis of 70-year-old pines in the Ruhr-District, Germany (after Wentzel, 1971). (a) Border zone, 10 km north of the Emscher River. (b) Central zone, on the Emscher River.

Growth comparison	(a)	(b)
Height of pine	19.7 m	6.8 m
Breast height diameter (bhd)	27.0 cm	13.8 cm
Timber mass/ha	401 m^3	38 m^3
Value/ha	20,050 DM	1520 DM

vehicles (Martin and Keller, 1974). Long known in California, where the damage caused by this type of pollution to the vegetation is estimated at 10—12 million dollars (Darley, 1969), they have now been confirmed in The Netherlands and Western Germany also (Knabe et al., 1973). This kind of pollution develops from a sun-induced reaction between nitratoxides and certain carbohydrates, mostly olefinic. Peroxyl-nitrates, ozone and nitrogen dioxide result from this reaction. These substances affect the processes of photosynthesis, respiration and enzyme formation. Visible symptoms of damage will be noticed only at concentrations of 4—8 p.p.m. (Darley, 1969);

106

they will look somewhat like the damage caused by sulphur dioxide. A survey of the phytotoxicity of nitrosulphuric gases is given by Bucher (1975). Investigations on the influence of ozone have been carried out by Wood and Coppolino, 1972, and Treshow and Stewart, 1973. Martin and Keller (1974) and also Keller (1974a) have shown that by investigation of the activity of the enzymic peroxidase, metabolic alterations caused by industrial or traffic pollution can be detected long before their external appearance. A comprehensive bibliography on air pollution and its effect on humans, plants and trees is given by Garber (1967, 1973). Individual problems with different kinds of pollution are described by Forstl. Bundesvers. anstalt, Wien (1971). Today lichens are frequently used as indicators for air pollution. They alert us to it in various ways over long periods (Skye, 1968; Ferry et al., 1973; Kirschbaum, 1972; Steubing and Klee, 1972; Thiele, 1974). Tree bark can also indicate air pollution (Grodzinska, 1971; Lötschert and Koehm, 1973a, b).

TABLE 37

Resistance of trees to sulphur dioxide

			Author
Very sensitive		Fir, Spruce Douglas fir	Wentzel, 1969
	Salix purporea	*Pinus sylvestris* *Larix decidua* *Picea abies*	Ranft and Daessler, 1970
Sensitive			
	Linden, Ash, Beech, Hornbeam Cherry, Plum	Pine, Larch White pine	Wentzel, 1969
	Berberis vulgaris *Salix fragilis* *Salix pentandra* *Tilia cordata*	*Pinus nigra*	Ranft and Daessler, 1970
Relatively insensitive			
	Oak, Alder, Poplar Maple, Elder Pear, Peach	Austrian pine *Arbor vitae* Yew	Wentzel, 1969
	Buxus sempervirens *Ligustrum vulgare* *Platanus acerifolia* *Quercus petraea*	*Juniperus sabina*	Ranft and Daessler, 1970

Wentzel (1963) deals with the possibilities and limits of forest protection against the hazards of smoke. Experts agree that there are no trees or other plants which are absolutely resistent to immissions. Sooner or later every plant reaches the point where it sickens or even dies. Resistance is always relative. It depends on:
(1) The type of immission (kind of gas), its intensity and period of influence;
(2) The plant's phase of development (age, season, general health condition);
(3) Growth conditions (soil, climate, nutrition);
(4) Location (distance from the ground, shielding by buildings or protective plantings).
Generally, deciduous trees and shrubs are more resistant than conifers. Young conifers are more resistant than older ones. Broad-leaved evergreens

TABLE 38

Resistance of trees to fluorine

			Author
Very sensitive			
	Beech, Hornbeam Linden, Peach	Larch, Spruce Fir, Douglas Fir	Wentzel, 1969
	Berberis vulgaris *Juglans regia* *Vitis vinifera*	*Larix decidua* *Picea abies* *Pinus sylvestris*	Daessler et al., 1972
Sensitive			
	Maple, Birch Ash, Elder Apple, Pear	Pine White pine	Wentzel, 1969
	Carpinus betulus *Rubus ideaus* *Tilia cordata*	*Pinus nigra*	Daessler et al., 1972
Relatively insensitive			
	Willow, Alder Oak, Red oak Locust	Australian pine Yew, *Arbor vitae* Juniper	Wentzel, 1969
Very insensitive			
	Acer campestre *Acer platanoides* *Euonymus europaeus* *Quercus robur* *Sambucus racemosa*	*Chamaecyparis* *pisifera*	Daessler et al., 1972

TABLE 39

Resistance of trees to nitrogen dioxide (van Hauten and Stratmann, 1967)

Very sensitive

White birch	*Larix europaea*
Apple, wild tree	*Larix leptolepis*
Pear, wild tree	

Sensitive

Acer platanoides	*Abies homolepis*
Acer palmatum	*Abies pectinata*
Tilia grandifolia	*Chamaecyparis lawsoniana*
Tilia parvifolia	*Picea alba*
	Picea homolepis

Relatively insensitive

Carpinus betulus	*Pinus austriaca*
Fagus sylvatica	*Pinus montana mughus*
Fagus sylvatica atropurpurea	*Taxus baccata*
Ginkgo biloba	
Robinia pseudacacia	
Sambucus nigra	
Quercus robur	
Ulmus montana	

TABLE 40

Resistance of trees to nitrogen trioxide (Ewert in Keller, 1973b)

Very sensitive

Alnus glutinosa	*Pinus strobus*
Alnus incana	
Carpinus betulus	
Tilia cordata	
Tilia tomentosa	

Sensitive

Acer pseudoplatanus	*Larix* species
Betula pendula	*Picea abies*
Fagus sylvatica	*Pinus sylvestris*
Fraxinus excelsior	*Thuja occidentalis*

Relatively insensitive

Acer campestre	*Chamaecyparis* species
Acer negundo	
Quercus borealis	
Quercus robur	
Robinia pseudacacia	

TABLE 41

Resistance of trees to ozone (Wood and Coppolino, 1972)

Sensitive	
	Green ash
	White ash
	Mountain ash
	Sweet gum
	Pin oak
	Scarlet oak
	White oak
	Hybrid poplar
	Sycamore
	Redbud
Relatively insensitive	
	European white birch
	Grey dogwood
	Flowering dogwood
	Little leaf linden
	Norway maple
	Sugar maple
	English oak
	Shingle oak
	Tulip poplar

are more resistant than those with narrow needles (Steinhuebel, 1961/62). All statements concerning the resistance of trees in the following specifications are therefore also relative (Tables 37—41).

De-icing salt harmful to trees. An alarming use is being made of de-icing salts to remove ice and snow from roadways. Most used is NaCl (95% of all applications); occasionally $CaCl_2$ is used (5% of all applications). Also, the excessive use of mineral fertilizers brings large quantities of salt into the soil and to plant roots. The salty water from melted snow and ice harms trees either by splashing on branches and shoots or by soaking into the ground to the root area (Plates 28 and 29). Because the toxic effect of de-icing salts was initially a point of controversy, many studies have been made so that today such toxicity is considered an established fact (Stach, 1969; Leh, 1972, 1973, 1974; Foss and Nehls, 1973; Ranwell et al., 1973; Glatzel, 1974; Kreutzer, 1974; Ruge, 1974; Glatzel and Krapfenbauer, 1975; and many more). Further literature in English: Davison, 1971; Eaton et al., 1971; Rundel, 1973; Smith, 1972; Thomas and Swoboda, 1970; Daniels, 1974. The salt causes damage to trees by osmotic absorption of water and by its specific ion effect on the protoplasm. Salt binds water; as the salt concentra-

110

Plate 28. Effect of de-icing salt (horse chestnut). Left: no salt spread; full foliage. Right: leaves falling as early as September (Photo Bernatzky).

tion increases, the water becomes less accessible to the plant. An NaCl solution of 0.5% binds water with a tow of 4.2 bar; one of 1% with a tow of 8.3 bar; one of 3% with a tow of 20 bar (Larcher, 1973a). Only plants with a high absorptive capacity can supply themselves sufficiently with water (Blum, 1974). Depending on the protoplasm's smaller or greater resistance to salt the latter causes greater or smaller metabolic impairment. Its tendency to swell impedes enzyme activity.

The toxic effect of the Cl^- ion on plants has not yet been completely explained. It is frequently maintained that toxic effects are not specific but results from the fact that during accumulation in the cell the Cl^- ion changes

Plate 29. Tree killed by salt (Photo Stummer).

the proportion of absorbed ions to free ions, in favour of the latter. This entails dehydration of the protoplasm and thus irreversible damage. Other authors are of the opinion that chloride accumulations impair the amino-acid and carbohydrate metabolisms (Leh, 1972). Sodium works in several ways. It impairs the metabolism of plants by disturbing the cation balance. Furthermore, it blocks the intake of essential nutrients (magnesium, calcium, potassium) by the roots, and their transport through the plant. Finally it contributes to deterioration of the soil structure as the Na ion is absorbed by the clay and humus particles, so that the other cations (Ca^{2+} and Mg^{2+}) are eliminated. This process destroys the crumbly structure of the soil; compaction sets in and the already adverse conditions for air and water in the soil decline further (Leh, 1972).

In the leaves of trees in Hamburg, Stach measured Cl-contents up to 6.3%

of dry weight. The absorbed salt will only partially be removed again by the fall of the leaves; the larger part of it will be stored in the wood of the trunk, in branches and in roots and from there will be transported back to the leaves in Spring with the transpiration stream. In the soil the salt accumulates more and more as a result of metabolic processes in the tree's ecological system (Fig. 36). The salt sprinkled in Winter is absorbed by the roots in Spring and transferred to the leaves. With premature or autumnal shedding of the leaves it gets back into the ground. During one vegetation period an additional amount of about 20—30 times the chloride contained in the litter will be washed out of the leaves by the rain, soak into the ground and be absorbed by the roots, and in this way again enter into circulation. Unfortunately, in most cases the washing out of the salt into the deeper layers of

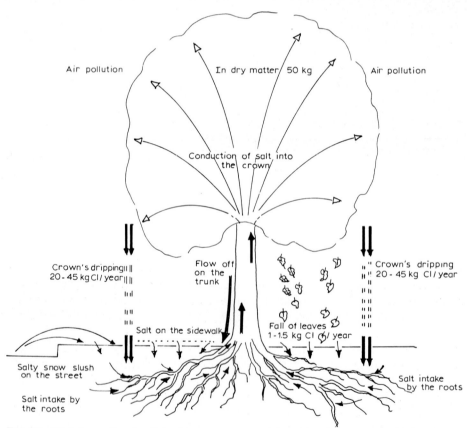

Fig. 36. De-icing salts circulate in street trees. After Blum, in an unpublished expert opinion of the Institute of Soil Science. (Salt quantities measured from Inst. Soil Science, Freiburg, W. Germany.)

the soil is not sufficient, depending as it does on soil structure and water conditions. If no more salt were to be used from now on, damaged trees would be able to recover completely only after 8—15 years (Hädrich and Blum, 1972).

Symptoms of salt damage on trees are: delayed sprouting of the leaves in Spring, leaves smaller than usual, necroses of leaf edges and blades, brown colouring and fall of the leaves, several new sproutings in Summer, more than one flowering during the year, drying of buds, drying up of whole branches and finally of the whole tree (Ruge, 1971a).

Possibilities for suitable action against salt damage are limited. The most promising solution is to grow and cultivate new species of trees and to select the ones most resistant to de-icing salt and its components, but this takes a long time. The possibility should also be investigated of developing thawing materials harmless to trees, but great caution is required since other suitable substances such as $MgCl_2$ and urea are also harmful to plants (Ruge, 1972a,b). Also, other thawing materials are far more expensive than NaCl. The only viable solution that remains is to improve the living conditions of the street trees as much as possible with a view to strengthening their resistance to salt and/or weakening the noxiousness of salt solution for plants. There are various ways in which this can be done.

First of all the water supply must be improved, especially for street trees. This requires the installation of effective irrigation systems and several good waterings during dry summers. Before sprouting in Spring, that is at the start of the sap rise, a thorough watering can help to wash out the salt or at least reduce it. The upper layers of the soil should be protected against drying out by suitable means such as covering the root area around the trunk of street trees with coarse gravel or with special porous slabs. This effectively prevents drying out of the living roots.

Most important is the improvement of soil aeration. This requires the removal of impenetrable layers over the root area, a measure which may, however, increase the danger of salt damage. Nevertheless trees do require a sufficiently large open root area for good development. Soil conditions may be improved by humus, which is extremely important for plant life. Sufficient and well-balanced nutrition is also essential (page 228). Chloride absorption will be reduced by an ample supply of NO_3 (Ruge, 1972a) and equally by phosphate ions (Leh, 1972). Conversely, NO_4^+ stimulates chloride absorption. Ruge particularly recommends K_2O, P_2O_5, Mn, and B. On no account should fertilizers containing Na_2O, CaO or Cl be used. The soil's calcium supply should receive careful attention since the formation of easily eroding $CaCl_2$ may cause calcium losses which impair the crumbly soil structure. An ample supply of potassium will reduce the absorption of sodium.

Another way would be to cultivate salt-resistant plants in soil with fairly high concentrations of NaCl to accustom them to their future street locations, thus preventing salt shocks at a later stage (Chrometzka, 1974a, b).

114

TABLE 42

Salt resistance of trees

Ruge, 1972a (after Walter et al., 1974)	Buschbom, 1972	Emschermann, 1973
Relatively tolerant		
Platanus acerifolia	Acer campestre	Acer platanoides
Quercus robur	Elaeagnus commutata	Fraxinus excelsior
Quercus rubra	Fraxinus ornus	Lonicera xylosteum
Sorbus	Halimodendron	Ribes alpinum
Crataegus	Lycium halimifolium	Rosa rugosa
Sophora	Populus canescens	Symphoricarpus albus
Robinia pseudacacia	Ribes aureum	Ulmus glabra
Fraxinus excelsior	Salix alba	
Tilia tomentosa	Tamarix species	
	Ulmus glabra	
Less tolerant		
	Hippophae rhamnoides	Acer campestre
	Alnus incana	Alnus glutinosa
	Lonicera xylosteum	Salix caprea
	Populus tremula	Ulmus carpinifolia
	Prunus avium	
	Prunus padus	
Very sensitive to salt		
Aesculus hippocastanum	Carpinus betulus	Carpinus betulus
Acer species	Betula pubescens	Cornus sanguinea
Tilia species	Cornus mas	Corylus avellana
	Cotoneaster integerrima	Crataegus monogyna
	Corylus avellana	Fagus sylvatica
	Fagus silvatica	Prunus serotina
	Picea abies	Rosa canina
	Pyracantha coccinea	Sambucus racemosa
	Prunus spinosa	
	Taxus baccata	

High borders will prevent salty solutions of melted ice and snow from being soaked up. This may be quite difficult to do with old trees, but with young plantings this precaution can be a big help (Fig. 91, page 245). In any case a reduction of the quantities of applied salt is possible because, according

Chrometzka et al., 1973	Daniels, 1974	Chrometzka, 1974b
		Decreasing salt compatibility
Elaeagnus angustifolia	*Acer negundo*	*Acer campestre*
Hippophae rhamnoides	*Elaeagnus angustifolia*	*Alnus glutionosa*
Viburnum lantana	*Fraxinus pennsylvanica*	*Alnus incana*
	Malus baccata	*Crataegus monogyna*
	Populus alba	*Crataegus oxyacantha*
	Morus species	*Robinia pseudacacia*
	Quercus alba	*Populus nigra*
	Quercus borealis	*Quercus robur*
	Quercus robur	*Quercus sessiliflora*
	Robinia pseudacacia	*Quercus rubra*
	Sensitive to salt	
Acer campestre	*Abies balsamea* [*]	*Acer platanoides*
Acer ginnala	*Acer saccharum*	*Salix caprea*
Acer pseudoplatanus	*Berberis thunbergii*	*Salix viridis*
Alnus glutinosa	*Buxus sempervirens*	*Betula pendula*
Alnus incana	*Carpinus betulus*	*Carpinus betulus*
Alnus viridis	*Euonymus alatus*	*Sorbus aucuparia*
Betula pendula	*Fagus grandiflora*	*Prunus padus*
Carpinus betulus	*Fagus sylvatica*	*Prunus serotina*
Crataegus monogyna	*Juniperus virginiana*	*Tilia cordata*
Crataegus oxyacantha	*Larix* species	*Corylus avellana*
	Malus species	*Sambucus nigra*
Corylus avellana	*Picea glauca*	Conifers
Ligustrum vulgare	*Picea pungens*	
Quercus rubra	*Populus nigra italica*	
Quercus multi-species	*Populus tremuloides*	
Salix caprea	*Pseudotsuga menziesii*	
Salix viridis	*Tilia cordata*	
Sorbus aucuparia	*Tsuga canadensis*	
Symphoricarpus orbiculata		
Symphoricarpus chenaultii	[*] *Acer pseudoplatanus*	
Prunus padus		
Prunus serotina		
Prunus spinosa		
Tilia cordata		
All conifers		

to expert opinion larger quantities than those given below do not yield better results. The quantity of salt applied must not exceed 40 g/m² or roadway in any case. Normally 15—25 g/m² is enough (1 teaspoon). With 40 applications in one winter this amounts to 1 kg/m². If 10% of it is absorbed by tree roots, and if all other precautions have been observed, this quantity could be endured by a well-nourished healthy tree. A strip of 1 m width along the kerb should be kept salt-free. The melting flow from the road will also thaw the snow on this strip. Only properly functioning mechanical sprinklers permit exact dosing. With shovels the quantities distributed are too large. In towns most trees grow on pavements, where thawing is not as urgent as on roads. Roughening materials like gravel, sand and chippings, etc., should be sufficient; when absolutely necessary, salt can be added in a mixture of 1 : 10. The areas below the tree crowns should always be kept salt-free, however. Salty slush should not be heaped onto the roots as is very often done on roads and streets. Furthermore, rapid drainage of melting snow should be made possible. If these recommendations are observed the benefit to trees will be considerable.

As the chloride ions penetrate the buds most easily after the leaves fall in Autumn, an attempt was made experimentally to prevent this by applying films of oil or synthetic material. These attempts were unsuccessful (Buschbom, 1972).

Salt resistance of trees. Not all trees are damaged to the same extent by de-icing salts. True salt resistance is a characteristic of protoplasm. With increasing chloride concentration of the soil, some plants develop xeromorphic qualities (Hayward and Long, 1941): thickening of the cuticle, succulence of the leaves, diminution of number and size of the vessels, increase of the osmotic value of the cell sap, stunted growth. Trees with very deep roots, receiving their water from deep and not very salty soil layers, will be less damaged (young plants are the exception). Some trees exude salt by excretion or guttation. Tamarisks, for instance, use glands or hairs for external secretion. There are many different opinions about the salt tolerance of trees. This is understandable since the sum of the ecological factors at work is of importance and laboratory tests can never take all components of local circumstances into account. Despite this uncertainty with regard to details, there also exists considerable agreement. Generally, conifers are more sensitive to salt than deciduous trees; of these the lateral-rooted are more sensitive than the deep-rooted ones. Most sensitive are apple, apricot, peach, plum, lemon and mulberry trees. Among street trees, the most sensitive are almost all species of linden and the horsechestnuts (Table 42).

Gas damage to trees. As early as 1954, Hoffmann reported damage to street trees caused by leaking city gas pipes. Today natural gas is used on a large scale and damage to street trees is evident everywhere (Kühne and Koester, 1967). The natural gas is drier and it is pumped through the pipes at higher pressure than manufactured gas. In this process, the old hemp packings of the pipe sleeves dry up and gas leaks out. In addition, the heavy vibrations caused by traffic loosen the pipe joints (Hoeks, 1972), (Table 43).

According to the formula $CH_4 + 2 O_2 \rightarrow CO_2 + 2 H_2O$ methane will be

117

TABLE 43

Composition of natural gas and manufactured gas in vol. percent (after Walter et al., 1974)

		Natural gas	Manufactured (illuminating) gas
Hydrogen		—	50 %
Carbon-monoxide	CO	—	7.0 %
Nitrogen	N_2	14.0 %	5.0 %
Methane [*]	CH_4	81.9 %	32.5 %
Ethane	C_2H_6	2.7 %	—
Propane	C_3H_8	0.4 %	—
Butane	C_4H_{10}	0.1 %	—
Pentane	C_5H_{12}	0.03 %	—
Higher carbohydrates	C_nH_n	0.04 %	3.5 %
Carbon dioxide [*]	CO_2	0.8 %	2.0 %

oxidized by certain soil bacteria into carbon dioxide and water (Adamse et al., 1972). In this way the bacteria extract from the soil air 2 molecules of oxygen per molecule of oxidized methane while they discharge carbon dioxide (Adams and Ellis, 1966; Adamse et al., 1972). This further deteriorates the aerial conditions of the soil in which street-trees grow, which is already compressed and sometimes covered with impenetrable material. The gas extends ball-shaped in all directions including the downward one. Measurements taken in The Netherlands showed soil air containing 6—2% oxygen and 10—15% carbon dioxide (Studiekommissie, 1970). The author's own observations showed that leaking gas can travel over long distances in the subterranean system of utility lines, such as telephone cables, until it finally escapes at places where no gas pipes exist. It is in these locations that trees die. According to reports of the Studiekommissie Invloed Aardgas op Beplantingen (1968—1972) up to 20% of street trees in The Netherlands are lost yearly through gas damage. Losses of 4 litres/hour/m^2 of pipeline surface have been found. Once a soil has been contaminated by natural gas, new trees should not be planted in it until several years have elapsed, that is, until the soil's oxygen content is back to 12—14%. According to Hoeks and Leegwater (1972), loose sandy soil permits the planting of trees immediately after repair of a gas leakage. Both authors found the contaminated soil to contain more nitrogen. Walter et al. (1974) found that the gas had increased the carbon content and thus that of humus. Remedy of gas damage naturally requires immediate repairs of the leaks, but until these are finished, as well as afterwards, it will help to blow air into the soil with a compressor at 7—10 atmospheres, to a depth of 0.60—1 m, for at least one hour. This has to be done very carefully to avoid an explosion of the gas—air mixture. Furthermore, holes may be drilled into the soil with an earth drill or an irrigation lance. Perforated pipes are put into these holes and at ground level they are covered with perforated slabs. However, to fill these pipes with gravel would hinder the air exchange. Hoeks (1971) found less than 10% oxygen in gravel-filled pipes, but 18—20% oxygen in empty ones. At sites of major damage, a whole series of ventilation shafts can be installed at distances of 50—60 cm. In addition to leakage repair and aeration by means of a compressor or ventilation shafts, damaged trees should also be fertilized and watered with due regard for species requirements.

4.4.2. Trees in physical planning.

Trees grow slowly, often requiring hundreds of years, but they can be cut down in a few minutes. The intention to replace them is hardly any consolation, nor is there any short-term gain if, in exchange for an old tree, a thin young sapling is planted. As shown on page 160, it takes 2700 young beech trees to replace a single, 100-year-old tree! Therefore it is reasonable to include existing trees in new plannings and preserve them as far as possible, provided they are healthy and still have a long life expectancy.

Planning of streets and roadways in densely populated areas often resolves around 'compulsory points' like bridges, hospitals, historical monuments and so forth. It is high time that we also begin to consider beautiful old trees as 'compulsory points' which planners must respect. Often this can easily be done by slightly adjusting the road line so that the area beneath the tree crown will not be used as roadway nor be covered with paving of any kind. As the construction work gets under way, the necessary protective arrangements require consultations with the contractor (page 238). Where existing roadways are to be widened it is undesirable that both rows of trees should be removed. Instead the widening should be done to one side, sacrificing one row only. Roadways should never be widened by reducing the distance between driving lane and tree row because living conditions for the trees would then deteriorate too much and the danger of accidents would be raised (Plate 30). If such widening cannot be avoided, the erection of rails and buffers is absolutely necessary for the protection of the trees as well as the traffic. Branches extending too far into the street can be marked by signs in order to avoid mutilation of the tree by their removal (Fig. 37). Occasionally residential areas are built right into existing forests. That leads to total destruction of the forest trees standing there. Even if some individual trees remain at first, experience has shown again and again that they will die later. The change in the ecological conditions under which they have grown up is too severe, the sun scalds the trunk, they easily fall victim to storms, the already small root area is further narrowed by construction work. The supply to the trees of soil, air, water, etc. is decisively changed. Forest trees can be kept alive only if they have been growing in a clearing for some time. Housing developments exist in forests at the expense of the trees and, alas, also of the people, who cannot endure the moist forest climate permanently.

Plate 30. The poor tree! (Photo Stummer).

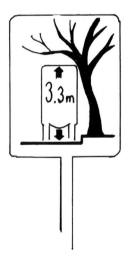

Fig. 37. Warning sign for low trees.

Tree preservation in physical planning in towns. Urban conditions are different when it comes to the construction of houses on tree-stocked sites. Here many trees can be preserved and secured (page 217 onwards). Infrared aerial photographs are recommended (remote sensing, page 273) to ascertain the number and health of existing trees and to accumulate the results in the tree register (page 277). Regular aerial photographs can also be helpful by showing trees otherwise hidden between buildings. The trees to be preserved are then marked on a special map giving the necessary information to the designers. Further protection of these trees depends on rules and regulations which are different in different countries.

Some countries and towns have special protective regulations for trees. In Persia, for instance, the unauthorized removal of a tree entails a fine and punishment by detention (Lechner-Knecht, 1974). In Vienna, Austria, the cutting down of a tree without permission by the authorities carries a fine of up to 500,000 shillings (about 27,000 dollars) or a 6-month prison sentence (Baumzeitung, 1975). Where good cause exists, permission will be given to cut down a tree but on condition that one substitute tree is planted per every 15 cm of trunk circumference, on the same site or as near by as possible. If this is impossible, then a compensation rate of 8000 shillings (about 440 dollars) per tree has to be paid to the authorities. This money is then used to plant trees in other parts of the town. Similar rules exist in The Netherlands, in Switzerland (Baumzeitung, 1972) and in Finland (Patoharju, 1974). Trees can be preserved even in difficult circumstances. The necessary measures are given on pages 217 and 233. Root treatment should be given special consideration. Because most injuries to trees result from construction work, road building and canalization, the authorities concerned should coordinate their activities well. The city of Hamburg, for instance, permits excavation in tree-stocked areas only on certain conditions which are set forth in the excavation permit and designed to prevent unnecessary tree damage (Baumzeitung, 1974). Some towns in Western Germany require not only that existing trees be preserved but also that new trees be planted by those acquiring a construction permit; the latter requirement differentiates between residential and industrial areas (Baumzeitung, 1971). According to these rules, shrubs must be planted on 10 per cent of the area not used for construction; the planting of one tree per 150—200 m^2 of unused area is also required. The authorities justify these rules by referring to the value of trees for the environment. Many more such regulations should be enacted by national legislatures. The Arboricultural Association of Great Britain has summarized the conditions for tree preservation in its leaflet No. 3 for designers (Hicks, without year).

4.4.3. Trees in front gardens
Front gardens have important functions for the buildings behind them, especially for the rooms on the ground floor. They act as dust filters and

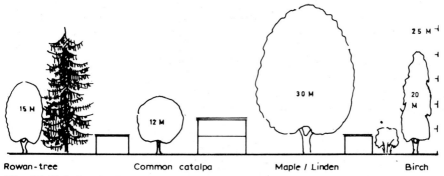

Fig. 38. Relation between house height and tree height. The height of the trees has to fit the height of the house.

noise screens. The temperature-reducing effect as result of heat consumption by evapotranspiration should not be underestimated (page 145). Designs of houses by architects almost always contain some smartly drawn trees in order to enhance the effect of the design. In harsh reality, however, the house stands all by itself — no trees far and wide. Perhaps this is out of fear that leaves will fall into the gutter and block it, or twigs damage the plaster of the walls. The first consideration is reasonable, but weighed against the beneficial effects of every tree, the chore of cleaning out the gutter every autumn is not worth mentioning. Damage to the house walls can be avoided

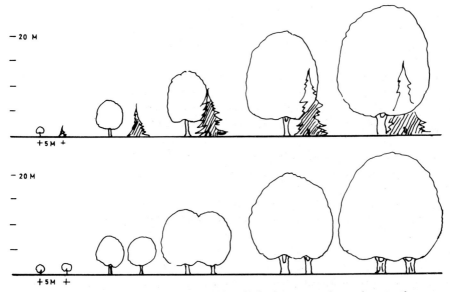

Fig. 39. Deciduous trees suppressing conifers if the latter stand too close to them.

122

Plates 31—33. Trees in front gardens (Frankfurt/Main) (Photos Bernatzky).

by selecting the right species for the available space and by planting it as far as possible from the house and towards the sidewalk. This gives the desired optical effect from the street area also, and life conditions for the tree are much better. Roots and branches can spread better, and a tree too close to the house will grow diagonally away from it into the space of the front garden.

Not every tree fits every house (Fig. 38). Where possible they should contrast with each other. If the house is low and flat the tree should and will surpass it. Where tree and house are of exactly the same size the proportions are not satisfactory. Should the front garden be narrow, pyramidal or columnar tree forms can be selected from various species (pyramidal oaks, narrow ginkgos, etc.). Deciduous trees suppress conifers standing too close to them (Fig. 39). To achieve satisfactory results all the house owners of a street should cooperate if possible. Where mutual agreements are not possible or are unsuccessful, the municipality should lend a helping hand. Plates 31—33 show how this is done in Frankfurt/Main where the location and species of the trees to be planted are laid down in the building plans (Bernatzky, 1969a). Trees in front gardens replace street trees, for which hardly any room is left nowadays because of the continuously increasing traffic.

4.4.4. Roof gardens

Growing construction and the resulting loss of green areas gives roof gardens some importance. They partially help to offset the deterioration of the urban climate even if they are not capable of replacing green areas and parks on the ground. At least they are better than nothing at all, and they are always psychologically effective in so far as occupants of the buildings can enjoy them.

The technical aspects of roof garden construction will not be discussed here. Because of the weight of moist soil, some special requirements have to be met (Gollwitzer and Wirsing, 1971; Gruen, 1972, 1975), (Fig. 40). Special consideration has to be given to irrigation and drainage, in case of heavy rainfall. For shrubs or small trees the soil height should be at least 45 cm. In order to keep the weight of the entire garden within reasonable limits this height should only be used in a few small areas. In large containers, trees or shrubs of medium height may succeed, but the ecological conditions are extreme in any case because roof gardens are an artificial and unfavourable location. Their effect on the urban climate is accordingly diminished. Generally all trees of arid climates are suitable; they are protected against excessive transpiration and may be planted if their root system is not too well developed. Trees and shrubs with a strong and extensive root system soon reach the limits of their potential. All plants used should be wind-resistant and capable of enduring quick drainage and repeated transplanting. They should also have a limited growth, controllable in size and weight. Nurseries still

124

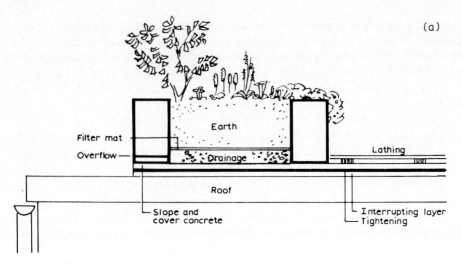

Fig. 40. Roof garden. (a) Profile; (b) view.

have much work to do in the cultivation of shrubs and trees which will not fail under the severe conditions of a roof garden. If the roof garden lies within the range of industrial emissions, or even just oil heating, care should be taken in using conifers and broadleaved evergreens. Large shrubs and trees must be secured against the wind. Hooks inserted into flagstones are useful for tying the steel cables used for anchoring the soil-ball of the trees. On ground level too, trees and shrubs may be planted in large containers of wood, stone or other materials (Fig. 41), but care must always be taken to provide good aeration of the soil, drainage and irrigation and, of course, fertilization as the quantity of soil is limited. To avoid heavy weights of earth on roof gardens, it is sometimes suggested that the soil may be partly

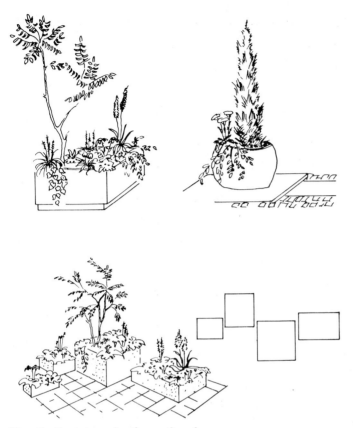

Fig. 41. Containers for the roof garden.

replaced by light foamy synthetics. It should be considered, though, that these are sterile materials and of no help whatever to the biological processes in the soil except perhaps for a limited water-holding capacity and aeration, which may also be achieved by a properly functioning irrigation system. On the other hand, light synthetics may be used instead of heavy gravel for drainage in containers and beds on a roof garden.

4.4.5. Trees for visual protection
Not everything in the human world is harmoniously arranged. Everywhere there is something to hide; unsightly sheds or warehouses, factories, sheds of intensive livestock-breeding, camping places with trailers and so on. The first step towards harmonious development and avoidance of disturbing scenes, lies in a satisfactory solution of the architecture of these buildings and facilities. Moreover, trees can do much to bring these buildings into a harmonious arrangement with the surrounding landscape. For this the planting

of rapid growing, high and broad trees as well as large shrubs of the native landscape will be the right thing in view of existing ecological possibilities. Exotic items usually fail in such locations because they will be unlikely to get the necessary special care, and the object to be hidden will only be more accentuated by them. Evergreen trees are particularly suitable for such purposes, but they should not be used where the ecological conditions do not suit them because then they would look just as strange as exotic trees. However, evergreens need not always be selected. The complex structure of a tree crown compensates for the vertical and horizontal lines of a building. Trees put the outline of a building into the background even if they do not cover it completely. This suffices to make the observer less conscious of them. Humans are often subject to powerful illusions regarding their observations. Metzger (1976) shows how to hide objects by applying the laws of perception.

The structure of such a planting for visual protection corresponds with that of any other protective planting. Both are also alike in that they usually receive little care and maintenance. Therefore trees and shrubs of a tough and unpretentious nature should be chosen. The combination of trees with shrubs of various heights will ensure a good cover. All trees and shrubs used should have a minimum height of about 2 m when planted. Experience has shown that smaller trees usually fail and die. Moreover they would take far too much time before they could serve their purpose. Visual protection should be achieved as quickly as possible and the planting must have twice the usual density. Some years later the surplus trees can be used elsewhere or cut down in order to avoid burdening the planting. Self-climbing woody plants like ivy or virginia creeper (*Parthenocissus*) have been used far too little up to now for covering up walls of buildings. The commonest objection is that they moisten the walls. The contrary is true. The overlapping leaves keep the rain off the wall, which cannot get wet, and the roots keep the soil dry near the foundations because the plant will absorb any available moisture and conduct it away by transpiration.

4.4.6. Age distribution of trees

Natural groups of trees and forest stocks show a multi-staged structure in terms of age and tree size. This makes for a lively appearance as well as stability. Only man created the artificial forest with trees of the same age; a structure which is neither satisfying aesthetically nor to be valued ecologically. Trees of the same age are felled at the same time which leads to clearing; ecologically a doubtful undertaking. In contrast, different ages in the tree stock make for a diversified picture that is also aesthetically satisfying. Planning a planting on this basis requires good knowledge of the life expectancies of the different tree species (Table 74). Here it becomes clear that life expectancies vary widely. A planting based on natural potential vegetation will automatically reveal tree stocks of varied life expectancies of

Plates 34 and 35. Trees in various stages of life (from Prince Pueckler) (Photos Bernatzky).

individual species. Almost every tree grows at its own speed, thus differing from its neighbour. This already diversifies the picture, thus preventing it from growing dull. However the structure and age development of a planting should not be left to chance. On the contrary, young undergrowth should be consciously planned and provided for, particularly where groups of trees of the same species occur. Where this is the case, contrasting old groups should be planned and preserved. This was carried out in an ideal way by the great 19th century landscape architect Fuerst Pueckler who created the famous parks of Branitz and Muskau in Germany (Plates 34 and 35). This brings us to the special discipline of the aesthetic aspect of the planning and lay-out of parks.

4.4.7. Domestic versus exotic trees

Whole books could be filled with the dispute about this. The question has many sides. First, a historical one. In Central Europe, before the last Ice Age, there existed many tree species which have since disappeared. The mountain ranges extending from east to west blocked their return after the ice melted. Can they be called exotic today? In North America, where the mountain ranges extend from north to south, no such question arose.

Of course important changes have meanwhile taken place in the various countries so far as ecological conditions are concerned. Hence the mere fact that certain trees existed in a particular country millions of years ago cannot by itself decide the question of whether they should be planted there today. The question of exotic versus domestic trees should be answered in some other way. We must differentiate between trees in the landscape and trees in enclosed gardens. Landscape plantings are expected to enhance the particularities and characteristics of their setting and to harmonize with it. The human enterprise must be in natural and aesthetic harmony with the landscape. This can be achieved only with domestic trees. Natural plant communities in their own habitat show a great harmony, no matter what their ecological condition. There is no danger of monotony. The quick change of the particularities of different habitats in the landscape always creates new pictures. The maps showing natural potential vegetation and plant communities (referred to on page 71 onwards) provide information about suitable tree species. Wherever these maps are not available a simple observation will help: almost everything growing vigorously by itself without much care fits ecologically and may be planted, but everything that does not grow vigorously should be abandoned, even if it is rare and expensive. Indigenous trees are far more resistant to pests and diseases than exotic ones, and they require little care. Therefore only indigenous trees should be planted in the open landscape. The city dweller, however, will often be unsatisfied with this. He wants to see conspicuous forms, weeping trees and pyramids as well as bright colours, so our parks are often dominated by exotic trees whose senseless mixture of form and colour produce the most glaring effects. In so doing, one acts against nature and actual human needs. We are continuously over-strained and over-stimulated in the daily requirements of our work, the stress of traffic and other demands of daily urban life. We do not need further stimulation from nature but rather soothing recreation. We need a proper balance and harmony instead of rush and activity. Natural harmonious plantings serve this purpose.

However, if exotic trees are not to be given up altogether they should be selected from related associations of ecologically adjoining areas. Then they will be the best warrant for a satisfactory development. In mixed plantings, where exotic trees are dispersed at random among indigenous species, they will cancel out each other's effect; a sure aesthetic touch as well as a good knowledge of ecological relations are therefore needed to place them correctly. A *Rosa centifolia* or a Chinese lilac in the middle of nature makes an unnatural and unappealing impression. Outlandish, multi-coloured and bizarre plants had best be planted in an enclosed garden whose owner may want to revive memories of journeys to foreign countries or who has some special dendrological interest and will plant exotic trees for this reason, but the difficulty of creating satisfactory natural pictures still remains. Even botanical gardens cannot always satisfy the observer's aesthetic requirements: they often give the impression of being a bad museum.

The public park, which is intermediate between landscape and private garden, will usually keep to domestic trees. If a park is large enough, exotic trees should not be dispersed among the domestic ones but planted together in groups by themselves. When this is done, their integration into the park will not be so difficult. The same considerations apply to conifers used in combination with deciduous trees. Wherever they are part of the natural vegetation they will not disturb the whole picture provided they are planted the way they grow naturally, but where they do not belong to the natural vegetation they should be handled with the same care as exotic trees. Their absence from the natural vegetation probably has ecological reasons.

In a book on ecology it is impossible to recommend certain trees or lists of trees suitable for all cases and usable in all surroundings. Trees for every site just do not exist! Every experience gained and every planting advice given is applicable only to a specific area.

5. THE CONTRIBUTION OF TREES TO THE ECOSYSTEM "TOWN"

5.1. Peculiarities of the urban climate

Trees and tree-covered areas, as well as parks and forests in the vicinity of towns are the best climate regulators. Their effect is best elucidated against the background of the individual components of the urban climate, the peculiarities of which become especially evident on hot and calm radiation days (heat relation-type weather), when the air is calm and the sky is cloudless, i.e. when no aerial exchange takes place. The urban climate differs essentially from that of rural land. For good comprehensive expositions on this subject see Kratzer (1956), Landsberg (1966), Chandler (1965) and others.

The differences between the urban climate and that of the surrounding rural areas are due to urban topography, buildings, the artificial supply of energy, the absence of vegetation and finally air pollution. These factors mainly affect and change
the intensity of solar radiation,
the temperature,
the relative humidity,
the local wind distribution,
the range of visibility,
the precipitation,
and other factors.
Various activities such as traffic, industry, domestic fuel combustion and so forth, continuously pollute the air in towns and above them (Table 44). These figures should be interpreted as relative rather than absolute. Average

TABLE 44

Number of nuclei (smaller than 5 μ) per cm^3 (after Landsberg in Kratzer, 1956)

	Average	Maximum	Minimum	Absolute maximum
City	147,000	379,000	49,000	4,000,000
Small town	34,000	114,000	5,900	400,000
Country	9,500	66,500	1,050	336,000
Coast	9,500	33,400	1,650	150,000
Ocean	940	4,680	840	39,800

TABLE 45

Concentration of trace constituents in a pure and in a polluted atmosphere (after Georgii, 1970b)

Trace component	Pure atmosphere	Polluted atmosphere
Particulate matter	$0.01 - 0.02$ mg/m^3	$0.07 - 0.7$ mg/m^3
Sulphur dioxide	$10^{-3} - 10^{-2}$ p.p.m.	$0.02 - 2$ p.p.m.
Carbon dioxide	$310 - 330$ p.p.m.	$350 - 700$ p.p.m.
Carbon monoxide	< 1 p.p.m.	$5 - 200$ p.p.m.
Nitrogen oxides	$10^{-3} - 10^{-2}$ p.p.m.	$10^{-2} - 10^{-1}$ p.p.m.
Total hydrocarbons	< 1 p.p.m.	$1 - 20$ p.p.m.

values as stated by Reifferscheid (1950) can be used as a basis for mathematical calculations: city 200,000 nuclei/cm^3; country 8,000 nuclei/cm^3.

Further observations have shown that aerosol concentrations in the atmosphere have greatly increased during the last few years. Davitaya (1971) reports that in some parts of the U.S.S.R. the dust content of the atmosphere has risen by 1.5—3 times during the last 30—40 years. He also informs us about measurements on Caucasian glaciers where the dust found during the Fifties was 20 times that found there in 1920. (Further statements by McCormick, 1967; Landsberg, 1970b.) Georgii (1970a, b) puts the aerosol proportion caused by industries at 30%. Poisonous waste gases adhere to these nuclei; CO$_2$, SO$_2$, nitrogen oxides, carbohydrates, etc. (Table 45).

Figure 42 presents information on the concentration of several gaseous

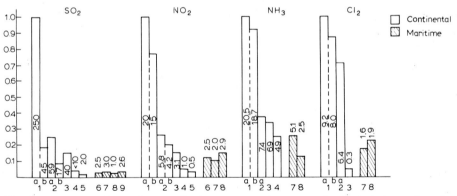

Fig. 42. Concentrations of atmospheric trace gases in pure and polluted atmosphere (after Georgii, 1970a). (Concentrations in μg/m^3).

1a Frankfurt M (Winter)
1b Frankfurt M (Summer)
2a Kl. Feldberg (Winter)
2b Kl. Feldberg (Summer)
3 Zugspitze (August)

4 St. Moritz (Summer)
5 Corviglia by St. Moritz (Summer)
6 Capreia Mediterranean Sea (Summer)
7 Florida
8 Hawaii
9 Meteor Expedition 1965 (40°—50°N, Dec.)

traces in pure atmosphere compared with the polluted atmosphere of the city of Frankfurt/Main. The concentration in Frankfurt during the winter was normalized to 1 and the concentrations found at the other locations are given as percentages of this value. The absolute values in $\mu g/m^3$ are indicated in the different bars. The significance of this pollution becomes evident when we realize that every person inhales at least 12 m³ of air daily and with it large quantities of noxious gases and poisonous substances (when hard at work a person inhales about 10 times more than the amount just mentioned). The harmful substances attached to the nuclei get into the lungs, and stay there or get transported into the blood stream. They become the cause of many diseases (Becker, 1971; Gutacker, 1972; see also page 167). This pollution causes a considerable change in the radiation budget above towns, as well as a reduction in sunshine hours. Chandler (1965) provides a table showing this reduction in London (Table 46). According to most authors the reduction of solar radiation amounts to 10—40%. Several authors emphasize particularly the reduction of the UV component of solar radiation, down to complete "UV-night".

Changes in radiation conditions above the town cause changes in temperature. The most striking feature of the urban climate is the temperature rise in town compared with that of the surrounding countryside. Depending on the size of a town, absolute temperature values may rise more than 10°C, but the average yearly rise will be 0.5—1.5°C. This is equivalent to a lowering of the town's altitude by about 100—300 m, with all the consequences this entails in terms of public health. Landsberg (1970b) sets the average temperature rise at 1.6°C; Dettweiler (in Landsberg, 1970b) cites 1.8°C for Paris. According to Chandler (1965) the difference in minimum temperature between the city of London and the suburbs may reach 6—7°C. Sekiguti (1970) concluded his investigations as follows. The heat island above the town is 3—4 times higher than the buildings; in Tokyo, the island is 100—150 m above ground level. The figures mentioned by Chandler and Clarke (given in Georgii, 1970b) — i.e., 50—150 m — are in agreement with these data. Addi-

TABLE 46

Averages of bright sunshine in London 1921—1950 (after Chandler, 1965, shortened)

	Hours per day		
	January	July	Year
Surrounding country	1.7	6.6	4.3
Outer suburbs	1.4	6.5	4.1
Inner high-level suburbs	1.3	6.3	4.0
Inner low-level suburbs	1.3	6.3	4.0
Central London	0.8	6.2	3.6

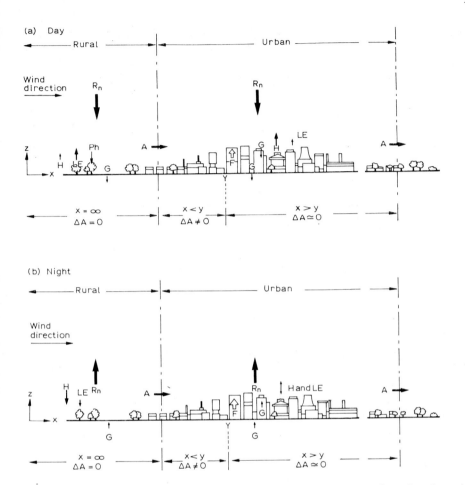

Fig. 43. Two-dimensional schemes of the heat balances of urban and rural surfaces (after Fuggle and Oke, 1970). R_n = net all-wave radiation; F = total heat artificially generated by city, due to energy released by combustion and metabolic processes; H = latent heat transfer; LE = convective sensible heat transfer; G = net heat storage in buildings, roads, ground, etc. ΔA = net advected energy.

tional causes for temperature divergencies between the city and the surrounding country are
heat accumulation in buildings and pavements,
artificial heat contribution from the combustion of various fuels,
impeded long-wave radiation due to increased pollution,
reduction of wind speed in towns,
lack of vegetation.
A town with its compact mass of buildings and streets practically resembles an artificial watertight rock. The surfaces of high buildings multiply the

effects of the sun's radiation many times; these accumulate a lot of heat during the day and radiate it off again in the evening and at night. This causes sticky "furnace heat" familiar to city dwellers, when the air does not cool off even at night. Thus town, constitute heat islands in the rural country. Concrete and asphalt possess a high thermic capacity and they are good heat conductors. The heat generated by combustion processes in the city and by animal metabolism may come close to, or even surpass, the amount developed by solar radiation (Fuggle, 1970; Fig. 43). The diagram represents an attempt to compare the urban and rural flows of energy, but only qualitative, not quantitative, statements may be made.

Analogous to the rise of temperature, is the considerable reduction in relative air humidity. At the beginning of this century, in a series of German cities this reduction amounted to about 6% (Kratzer, 1956); Bernatzky (1960) found values of 5—15% in Frankfurt. These facts explain the large increase of diseases of the throat, nose and ears.

Wind decrease — In almost all cities, meteorological stations report an increase of calm days and wind-speed reduction. Buildings cause a roughening of the ground surface which in turn causes a diminution of wind speed because rubbing on the walls uses up part of the wind's energy. "Therefore the aerial circulation is slowed down in the center of the town and immediately in front and above it. An air pillow is formed, onto which the incoming airflow must rise" (Kratzer, 1956; also Schmidt-Burbach, 1973; Munn, 1970). The amount of friction is greater in town than in the forest. Up to a height of 100 m above ground it surpasses that of the open country considerably. Sometimes the wind in the streets touches the ground, causing the aerial movement to appear stronger there than in the open country, so that stormy conditions bring violent gusts from a "false" direction causing unexpected damage. However, these turbulences produce no ventilation, for the air circulates in a confined space, so that exchange takes place only between higher and lower levels without any air being conducted away. Indeed the whirled-up dust may add to the pollution. The air flows in the streets vary with the latter's orientation and a circulation develops from cool, shaded walls towards sunny ones (Schueep, 1974). The alteration of the wind field, particularly in the area near the ground, has considerable consequences for the amount of waste gases in the air. Air whirls at the foot of skyscrapers in the luff, and turbulence zones in the lee, have been observed in the wind channel; these seriously hamper vertical air exchange (Zweiacker, 1975). Children's playgrounds or recreation areas in such positions therefore have the worst air conditions. Wind decrease in towns finally entails a reduction in cooling and thus of the physiological oxidative processes of the inhabitants.

On the other hand on calm days a special wind system develops, the so-called "field-wind" ("Flurwind"). As a result of the heating up of the conglomeration of buildings the hot air rises; at the same time, air flows from the outskirts to the town centre. This entails severe pollution of the inner city because the incoming wind picks up dust and waste gases on the way. That is why we find the worst air in the inner city (Fig. 56a). The wind distribution in radiation weather in Frankfurt/Main (Georgii, 1970a) confirms the above scheme. Onstruction of the long-wave radiation by the town's haze is to be noted.

The absence of vegetation greatly contributes to the process of overheating of the inner city. Vegetation consumes at least a small part of the sun's radiation by photosynthesis, but through evapotranspiration a large amount of radiation energy is consumed. Baumgartner (1956) found that above a young spruce forest about 66% of radiation energy was taken up by evapotranspiration during the day. Under extreme conditions it could be up to 90%. However, in a town the ground is hermetically sealed by waterproof and air-tight pavements. Precipitation cannot soak into the ground any more. It is drained off rapidly into the sewer and so withdrawn from evaporation. If we estimate the drained-off quantity of rain at 1/3 of the 600 mm average in Central Europe it would come to 200 litres/m^2/year. Since the evaporation of 1 litre of water requires about 600 kcal, it follows that urban consumption is 120,000 kcal behind that of the open country. This fact is noticeable as heat. The factors causing the considerable overheating of the town climate result in a non-stimulating, indulgent climate and an increased susceptibility to colds, headaches, aversion to work, etc. Details are treated on page 164 onwards. Against these negative consequences of a denatured climate in towns, trees and green areas planted with trees produce a number of powerful effects.

5.2. Effects of trees

After the pollution-reducing effect of trees and forests had already been investigated in the thirties (Loebner, 1935), increasing pollution prompted further research regarding the effects of vegetation (Foy, 1970; Garber, 1967, 1973; and others).

5.2.1. Reduction of particulate pollution

In judging the reduction by vegetation of particulate pollution we must distinguish between
dust reduction *on* vegetation areas, and
dust reduction *caused by* plantations.
With few exceptions (pollen), plantations themselves do not make any dust. Dust drifting in from the outside is mostly held by the vegetation (Hennebo, 1955). As a result of the wind reduction effected by the plantings, impurities in the air settle because the air's carrying capacity is largely reduced (Kreutz, 1952, Kreutz and Walter, 1960). Open plantings, where the wind can get

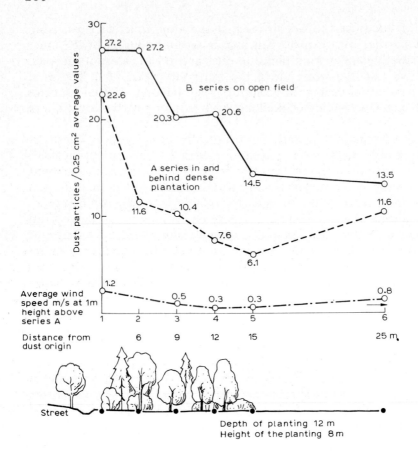

Fig. 44. Dust sedimentation on a dense garden planting and on an open field (Hennebo, 1955).

through, show a different reaction than dense ones. In the latter, dust values diminish rapidly inside the plantation, reaching high values on the luff side followed by a minimum on the lee side. However, they then increase again, while the dust values of comparative measurements above the open country diminish at a relatively steady rate, depending on the distance from the source of dust (Hennebo, 1955; Figs. 44 and 45). Scattered open plantings have their highest dust values immediately behind the plantation and from there the values diminish continuously with the distance from the source of dust. Dense plantings are swept over by increased wind velocities and the light nuclei are carried along over the obstacle. In scattered, wind-blown plantings, however, the incoming air-flow settles in the plantation or just behind it. In dense plantations also a fall-out of air pollution occurs as a side effect of turbulences, though not to the same degree as in loose plantings (Bernatzky, 1968). The smaller the nuclei the less they are filtered out; instead, they are

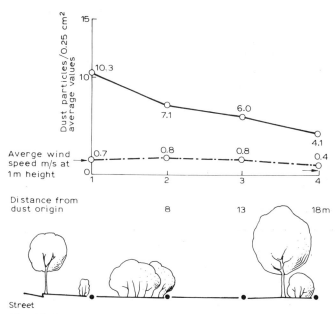

Fig. 45. Dust sedimentation on a loose field hedge.

carried away by the air flow after passing the plantation. The results shown in Table 47 of measurements in Frankfurt/Main on a calm radiation day, may serve to indicate the dust-clearing effect of plantations. Graefe and Schuetze (1966) measured, in the nearly treeless city of Hamburg, a dust fall-out of more than 850 mg/m²/day as a yearly average, but in some wooded areas in the outskirts as well as in all city parks measuring several hectars, it was less than 100 mg/m²/day.

The dust-filtering effect depends on the physical, chemical and physiological properties of the species of tree (Steubing and Klee, 1970). Loose hedges of *Carpinus betulus* showed a larger sedimentation of dust in the apical region of the plants; *Picea abies* more in the central and basal regions. *Pinus*

TABLE 47

Particulate pollution per 1 litre of air in a vegetationless city, in a park and on tree-lined streets (after Lamp, 1947)

	Particles per 1 litre of air				
	Morning	Noon	Evening	in %	%
Central station	16,830	18,310	17,640	100	
City centre	15,120	13,220	18,370	88.5	
Treeless street	12,880	10,180	11,490	65.5	100
Tree-lined street	3,870	3,040	3,830	20.3	31
Park	3,260	1,180	3,140	14.4	22

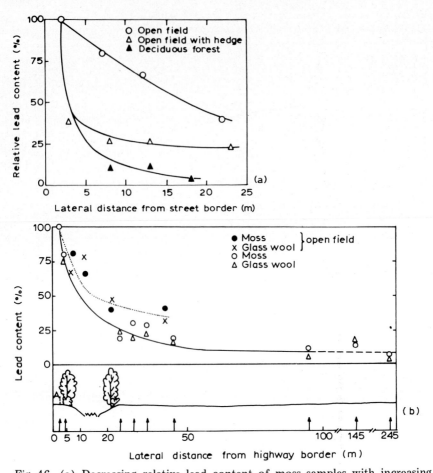

Fig. 46. (a) Decreasing relative lead content of moss samples with increasing distance from street border (after Keller, 1974c).
(b) Influence of a small field wood along the highway on the lead content of moss and glass wool samples (after Keller, 1974c). A = highway-border; arrows = measurement points.

mugo showed better screening than *Rhododendron catawbiense*. Dochinger (1972) confirmed the filter effect of trees, judging conifers to be better than deciduous trees. Flemming (1972a) dealt critically with the problem of dust filtration by trees. He discusses qualitative models of dust protection effect which are needed for the interpretation of measurement results and for more effectiveness in practical service. Keller (1974a, b; here also further references about the lead problem) has investigated the effect of lead contamination on woods, and the effect of the presence of trees on the level of lead contamination. He ascertained that protective plantings and forests considerably diminish the lead content of the air (Figs. 46 and 47). In the discussion

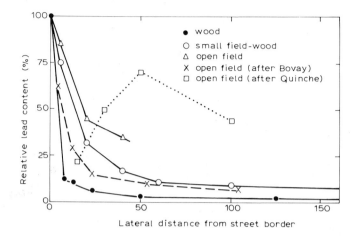

Fig. 47. Relative lead content of trees in relation to lateral distance from street border under heavy traffic (after Keller, 1974d).

about the result of his investigation, Keller energetically takes issue with Flemming (1972a) and Lampadius (1963) in defending his opinion that lead reduction is an effect of protective planting. Numerous publications exist on the filtering effects of the forest (e.g. Blum, 1965; Neuwirth, 1965; Wentzel, 1960, 1969; Keller, 1971a, 1973b; Knabe, 1973a,b; Braun, 1974). These effects consist of

active filtration (sedimentation, active adsorption and absorption on and by surfaces,
protection of sedimented pollutants against pick-up by the wind,
passive diversion of the air flow and resulting turbulences including sedimentation.
The extent of the filter effect depends on
the position of the forest in relation to the object to be protected,
the kind and concentration of pollutants,
the structure of the forest, particularly the degree of resistance of the different tree species,
the meteorological conditions,
the topography.
Meldau (1956) estimates the possible amount of bound dust on spruces at 32 t/ha, and on beeches in a wood up to 68 t/ha. That means that in an extreme case a forest is able to bind several times its own weight in dust upon its leaves and needles. Scientific soil investigations show exactly the dust quantities and qualities washed off by rain into the ground (Enderlein and Stein, 1964). Mayer and Ulrich (1974) calculated the considerable filter effect of a beech wood in the Solling (Germany) as regards air pollution (SO_2 and aerosols).

5.2.2. Reduction of gaseous pollution

Here things are not quite as clear. We know that gases are able to damage trees or kill them (page 103 onwards). On the other hand, positive results exist. Martin and Barber (1971) report loss of atmospheric sulphur dioxide near foliage. Ruge (1972b) suggested that gases in sublethal concentrations (SO_2, CO, nitrogen oxides) may be neutralized by oxidation through the plant metabolism. As mentioned before, noxious gases accumulating into particulate matter may, together with the latter, be filtered out by the trees. Roberts (1971) refers to American literature on this subject, but otherwise the protection against gases is judged very critically (Weidensaul, 1973). Lampadius (1963, 1968) declares that specific research on the filter effect of forests on gaseous pollution has not yet been done. According to him, the deciduous forest has no perceptible SO_2 filter effect as compared with the coniferous forest. Materna (1963) presents an analogous opinion. According to Lampadius, forest stands do not distinctly prevent the lasting penetration of immissions into the interior. Indeed in the centre of the wood diminution of high SO_2-concentrations takes more time than in the open. However, he considers protective plantings indispensable for fending off damaging climatic elements which, as experience has shown, may noticeably raise the disposition of forest trees against SO_2-damage. Further details about the problem of reduction of smoke concentrations by forest tracts are mentioned by Flemming (1967a). He distinguishes four effects: reduced wind speed; raised turbulence, its effect extending beyond the lee edge; true filtration by the plants — mostly slight; and physiological relief of the plants as a result of wind protection. According to Flemming, true filtration, that is gas deposits on plant tissue, exists but is often over estimated. He believes true filtration to be limited, especially that of SO_2. With other gases it might be different. Wind protection has the strongest effect in regard to smoke protection. For plants, excessive wind speeds are not favourable because they increase the sensitivity to smoke.

The influence of plantation screens on the dispersion and ground deposition of atmospheric pollutants is dealt with by Belot and Caput (1973). Hence the influence of foliage screens and the dispersion of atmospheric pollutants depends, to a large extent, on the nature of the plants in the screen. Thin screens, such as rows of trees, take effect by diverting or redirecting moving air currents. The thicker screens, such as clusters of trees or heavily wooded areas, tend to filter the masses of air and accumulate the particulate or gaseous pollutants on their foliage. In their report the authors present and discuss details on the effects of foliage screens on air currents and the absorption of various pollutants, particularly radionucleides. On the other hand, it is possible that vegetation increases the roughness of the ground and reduces the wind so that pollution cannot flow off. Much depends on the interaction between topography, meteorology and plants.

5.2.3. Shelter belts (protective plantation screens) (Plate 36)

The filter effect might be used for the lay-out and planning of plantation screens (Bernatzky, 1968, 1969c). The smaller and lighter the aerosol nuclei are, the better will they be carried in the air flow which takes them across the screen barriers. Therefore, to obtain a distinct effect at long range the screen barriers should be laid out at regular intervals, preferably at a distance of about 20—25 times their height. This would effectively block the impurities in the air by screening out coarse dust and turning away the nuclei through continuous re-elevation and diversion of the air flow from the object of protection (Figs. 48—50). As comparable accurate measurements on such projects are lacking so far — the difficulties lie in the diversities of the initial conditions, meteorological data, topographic requirements, etc. — adjusting wind protection measurements in open land (Naegeli, 1941, 1943, 1954, 1965) and in the wind channel (Kreutz and Walter, 1960; Caborn, 1965) must be referred to. They show, as a rule, that diminution of wind speed by at least 10% is achieved in front of the plantation at a height about five times that of the plantation, and behind it at a height about twenty five times that of the plantation. Best results are achieved when the screen planting has a

Plate 36. Natural hedge country in France (near Autun) (Photo Bernatzky).

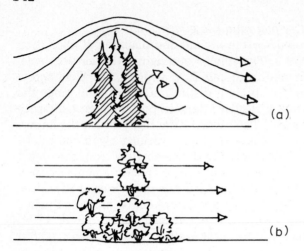

Fig. 48. Protective effect of various plantations. (a) Dense plantation: small filtering effect; whirls. (b) Loose plantation: blowing through; good filtering effect.

wind permeability of 40%. In Winter, in the leafless stage, deciduous plantations have about 60% of the summer effect. The observations made in the wind channel correspond well with investigation results in open land. Moreover they show that too dense a formation of the screens does not bring any greater advantage. Results of these investigations apply to all plantation screens, though naturally local variations must be taken into account.

To sum up, shelter belts act as a dust filter especially when laid out at right angles to the main wind direction. Depending on situation and purpose,

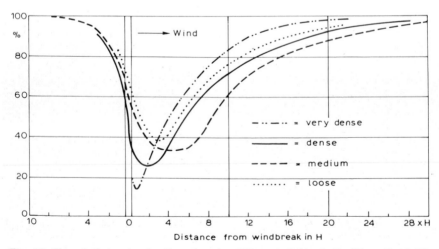

Fig. 49. The wind speed reduction given by different shelter-belts (Naegeli, 1943).

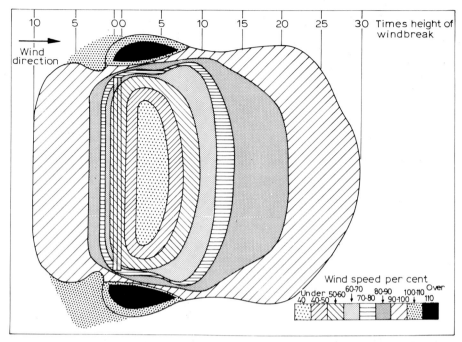

Fig. 50. Plan of wind conditions in the vicinity of a moderately dense shelter-belt. Wind speed expressed as a percentage of the free, unobstructed wind (after Caborn, 1965).

open and permeable screens may be combined with those planted to obstruct flows, particularly of gases. Concentric plantation rings around areas to be sheltered, combined with radial green junctions, screen the enclosed areas on all sides. However, the shelter belts around industrial areas can by no means replace the technical purification of industrial emissions; instead, they serve to support the latter, particularly since the emission residues are not eliminated by technical means. Shelter belts are suitable as wind breaks in agriculture and generally in the open country, where they help to increase agricultural production (Kreutz, 1952; Geiger, 1961; van Eimern et al., 1964; Naegeli, 1965; Caborn, 1965; and others). Behind their shield, the average production increase is more than 10%. Wind decrease reduces the evaporation and raises the CO_2 percentage of the air with a positive effect on photosynthesis and thus on production. Shelter belts also prevent erosion by wind and water. They always require intensive care (Haverbeke, 1973; Slabaugh, 1974). In areas with plenty of snow, shelter belts may direct the snow drifts. Solitary trees often cause long and high snow cornices. Dense plantations also cause snow drifts (sedimentations), but scattered, open ones allow the snow to be distributed uniformly (Mazek-Fialla, 1974). The planning of such snow plantation screens depends on the objective to be accomplished (Fig. 51).

144

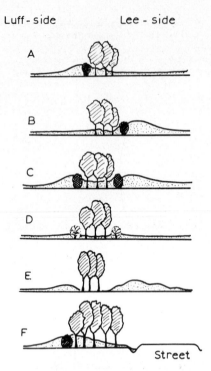

Luff-side   Lee-side

A

B

C

D

E

F

Street

Fig. 51. Influence of tree plantings on snow accumulation (Mazek-Fialla, 1974). A = dense shrubs on the leff-side; B = dense shrubs on the lee-side; C = dense shrubs on either side of the planting; D = loose shrubs on either side of the planting; E = snowbank alongside the planting without shrubby under-growth; F = planting along streets.

5.2.4. Protection against radioactivity

Large stands of trees and forests may reduce radioactivity in the air. Although they do not destroy it, they do alter its dispersion (Herbst, 1965, 1968). With particulate radioactive matter in the air, the foliage of investigated trees showed up to four times as much total radioactivity on the luff side as on the lee side. Fodder and food crops on the lee side of forests showed only 1/5 as much radioactive contamination as plants on the luff side. In individual cases the level was as low as 1/20 of the general contamination. Even in urban areas with little greenery, only 60—70% of the fall-out measured in places without any plants, was found. With a content of 1 microcurie/cm^3 volatile radio-iodine in the air (^{131}I) it was found that 1 kg of foliage took up as much as 1 curie radio-iodine in one hour at medium airflow. (1 curie radio-iodine corresponds approximately to the radioactivity of 1 g of radium.) Two-thirds of the radio-iodine was attached to leaf surfaces and could partially be washed off. One third entered the leaf tissue, probably via the stomata. Herbst concluded from this that even a reduction of 30—60%

of incoming radioactive material, which is possible under the shelter of forests, may help to preserve health and life. "A radiation burden of the whole body with 600 radiation doses (rad) brings a mortality rate of almost 100%, a diminution of the burden by only 1/3 would reduce this to nearly 50%, and 2/3 reduction would bring the mortality rate to almost zero".

5.2.5. Reduction of temperature

A short walk out of densely built up city areas into an adjoining park or even a tree-lined street is enough to show the cooling effect of trees and vegetation. For a long time no exact investigations existed. More recently, several detailed examinations have been carried out, allowing exact statements to be made about the air-cooling effects of the vegetation.

During one vegetation period, Bernatzky (1960) investigated in Frankfurt/ Main a ring-shaped green area encircling the city; its length was 5 km and its width 50—100 m. This green ring covered the former site of the old medieval fortifications around the city; it was planted with trees at the beginning of the nineteenth century. Along the length of this green ring, by means of five measuring stations 1 km apart, a longitudinal and several lateral profiles were produced; here, temperature, relative humidity and vapour pressure were measured. Each of the three measurement series, done three times daily, started and ended at a control measuring station in an open square in the town centre. The longitudinal profiles clearly showed lower temperatures in the green ring; these were especially evident on calm radiation days marked by the presence of a zonal high pressure bridge. On these days, the more the air temperature surpassed 30°C, the more conspicuous was the cooling effect of the green ring. On two days it surpassed 3°C (at a maximum air temperature value of 30°C). Distinctly clear was the regular lowering of temperatures towards the evening with increased contrast between city and green area despite already falling city temperatures in the same hours. The most conspicious temperature drops were observed at noon and also in the evening at the narrowest places of the green belt under the foliage canopy of 100-year-old plane trees, whereas on larger lawn areas the smallest falls were observed. This cooling effect is noteworthy because it was observed on a narrow green belt in the middle of a densely built up city. The stronger the wind blew and the lower the air temperatures, the smaller (as expected) were the temperature differences between the town centre and the green belt (Table 48). The biggest temperature drop measured was 3.5°C. Assuming an air temperature drop of 1°C with every 200 m of increasing altitude, the 3.5°C decrease corresponds with an elevation of 700 m. This means that the temperature of the green belt around Frankfurt city equals that on the 700 m higher Feldberg; a peak in the Taunus mountains about 30 km from the city centre (Fig. 52). This cooling effect is due not only to the diminution of radiation by the leaf canopy of the trees, as is often assumed, but also to the reversal of the facts described in the section on the causes of heat islands. The main point is

TABLE 48

Medium values of the lowering of temperatures (after Bernatzky, 1960) in centigrade

Time	27.6	28.6	30.6	7.7	8.7	10.7	11.7	14.7	15.7	21.7	19.8
8 — 9	—	0.20	0.56	0.78	0.52	0.28	0.32	0.72	0.76	0.62	0.14
12 — 13	0.68	1.06	1.90	1.22	1.28	0.48	0.72	0.76	0.12	1.16	1.42
20 — 21	0.84	1.50	2.60	1.42	—	1.36	—	0.82	0.82	0.88	0.21
Daily average	0.76	0.92	1.69	1.14	0.90	0.71	0.52	0.77	0.57	0.89	0.59

TABLE 49

Medium values of the rise of relative air humidity (after Bernatzky, 1960) in %

Time	27.6	28.6	30.6	7.7	8.7	10.7	11.7	14.7	15.7	21.7	19.8
8 — 9	—	3.2	1.6	3.4	3.4	5.0	2.0	4.0	4.0	4.6	0.2
12 — 13	3.2	3.6	4.4	5.4	1.4	4.2	2.2	3.6	3.6	1.4	9.4
20 — 21	5.0	7.6	7.0	4.6	—	12.8	—	3.6	3.6	3.6	1.6
Daily average	4.1	4.8	4.3	4.5	2.4	7.3	2.1	3.7	3.7	3.2	3.6

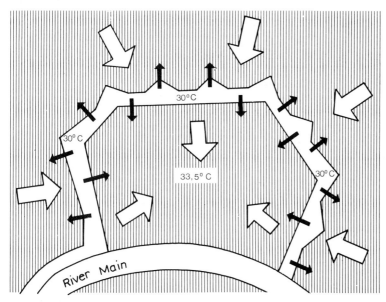

Fig. 52. Effect of a green belt on temperature in Frankfurt/M/W. Germany (after Bernatzky, 1960). White arrows, increasing pollution and heating up. Black arrows, supply of cleaned and cooled fresh air.

evaporation. It uses an average of 60—75% of radiation energy. For the cooling effect to materialize the radiation used by evaporation must be conducted away again.

The rise of relative humidity in the air. The lowering of temperatures corresponds with a rise in relative humidity, which is always too low in towns. In these measurements, an increase of 18% was registered which did not surpass the comfort limit nor even come near it. The increase was, like the rise in temperature, larger in the evening than at other times during the day (Table 49). These results were confirmed by Brahe (1974, 1975). He investigated the daily radiation balances and the air temperatures in treeless and tree-covered places as well as the relative humidity of the air or the vapour pressure (Figs. 53, 54). His conclusion is that tree-covered areas have a year-round cooling effect; in Winter on account of relatively high radiation at generally low temperatures; in Summer on account of low radiation at generally high temperatures. To eliminate the radiation surplus in towns he suggests laying out as many evaporating vegetation areas as possible, especially near the town centre. In order to carry off the surplus energy, the air exchange with the surrounding country should be supported by vegetation aisles. To support convectional air exchange, streets should be planted with trees on one side only preferably that facing south or west. Important air

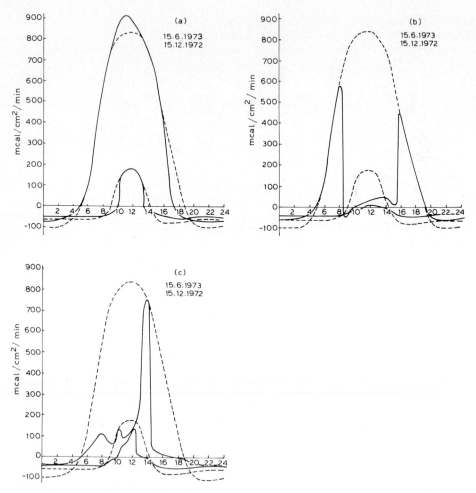

Fig. 53. Daily course of radiation balance in Aachen (W. Germany) (after Brahe, 1974). (a) In front of south wall (measurement point 3); (b) in front of north wall (measurement point 15); (c) under trees (measurement point 9). Drawn line, real daily course; dotted line, theoretical radiation balance in the absence of a surmounting horizon. The upper curves are for 15 June 1973; the lower ones for 15 December 1972. Total daily radiation balance (cal/cm²/d)

Measurement point	Date	Incident radiation	Emitted radiation	Daily total	% of theoretical balance
3	15.12.72	30.8	−62.8	−32.0	117
	15. 6.73	427.2	−38.0	389.2	103
15	15.12.72	—	−60.4	−60.4	222
	15. 6.73	122.8	−36.0	86.8	23
9	15.12.72	11.6	−55.6	−44.0	162
	15. 6.73	104.4	−23.2	81.2	21.5

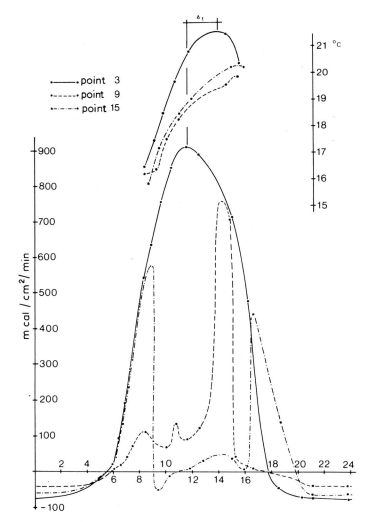

Fig. 54. Radiation balance (below) and air temperature (above) on a radiation day in Aachen (Brahe, 1974). Point 3, in front of a south wall; point 9, under trees; point 15, in front of a north wall.

aisles should not be blocked off by too dense planting. Sperber (1974) investigated green places of various sizes in Bonn (W. Germany). Brahe's observations about the vapour pressure in motor cars in parking spaces with and without trees, are interesting. In shadeless parking areas the vapour pressure in cars rose so much as to justify fear of heat accumulation, which impairs driving capacity and so increases the likelihood of accidents.

Ruge (1972b) calculated for Hamburg that of an annual average of 771 mm of precipitation, about 1/3 is led off through the sewers without evap-

TABLE 50

Some comparative micro-meteorological observations in a courtyard near grassland and woodgrove (summer) (after Landsberg, 1970a) *

Time	Air Temperatures			Humidities			Surface Temperatures							Remarks
							Walls facing				Ground			
	T_c	T_g	T_w	H_c	H_g	H_w	N	E	S	W	p	g	w	

Case 1 — Radiative cooling

LST 1620	30.6	30.6	30.0	50	52	52	32.0	35	34.5	> 44	> 44	33	30	3/10 cu, cs, sunshine, u = 3 m/sec.
1934	28.3	27.8	28.3	56	56	57	30.5	31	31.5	33.5	33	29	27.5	4/10 ci, cs (sunset 1915) u = 1 m/sec.
2115	25.6	24.7	24.4	65	65	66	27.5	28	29.5	29.5	30.5	23	25	1/10 ci, dew on grass u < 1/2 m/sec.

Case 2 — Cooling by thunderstorm

LST 1400	32.8	32.8	31.7	34	36	39	33	39	>44	>44	>44	36	31.5	7/10 cu cong, sunshine u = 6 m/sec.
1800	27.8	27.8	26.7	41	43	45	31	32	34	34.5	37	26	28	1/10 cb, begin thunderstorm u = 4 m/sec.
2130	24.5	23.3	23.9	68	67	89	25	27	26.5	28	29	21	21.5	10/10 st, sunset 1922 forest soil and grass wet (6.5 mm precip. 1825—1915); parking lot dry, calm

* All temperatures T in °C; relative humidity H in %. Symbols: C, courtyard; g, grass area; w, wood lot; p, paved parking area in courtyard. For measuring the ground temperatures the infrared thermometer subtended an area of about 25 cm², for the wall temperatures about 1000 cm².

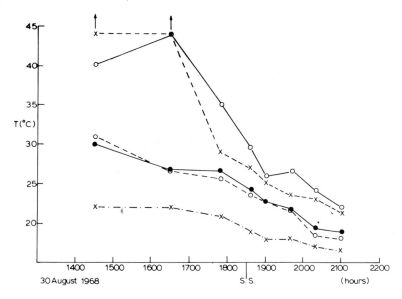

Fig. 55. Cooling on a clear night in a continental air mass. Note relatively rapid radiative cooling already starting before sunset. Air in courtyard again stays warmer than air over grass (after Landsberg, 1970a). o———o, west-exposed wall; ×— — —×, parking lot pavement; ●———●, courtyard air; o— — —o, air over grass; ×— · — · —×, grass surface.

orating; that is 257 mm = 257 litres/m^2 or 2570 m^3/ha. Since, according to him, a street tree evaporates 5 m^3 of water yearly, 500 street trees/ha would evaporate the same quantity of water. This would result in a cooling effect of 1.5×10^9 kcal/ha/year. When trees are lacking, that quantity of energy stays in the town. Landsberg (1970a) investigated conditions in a paved courtyard, 32 × 42 m in size surrounded by 18 m high buildings. On one side there was a widened roadway with parking spaces. On three sides a broad lawn area surrounded the buildings. In addition, at a distance of 150 m, there was a deciduous wood. These measurements clearly showed the cooling effect of wood and lawn as against the heat accumulation of buildings and pavement (Table 50; Fig. 55). On a larger scale the thermic difference between built up areas and open land spaces were investigated by means of thermal photographic measuring flights (RPU, 1972). This method is particularly suited for the registration of larger green areas and forests and their temperature-lowering effect, as well as of areas with the strongest cooling effects (Voelger, 1972; Lamp, 1972; Lorenz, 1972).

5.2.6. Influence of trees and green spaces on urban ventilation

The field wind developing on hot, calm radiation days above the town can be an effective aid in urban ventilation. In a city without trees and green areas the wind will continuously pick up pollution particles, but trees and

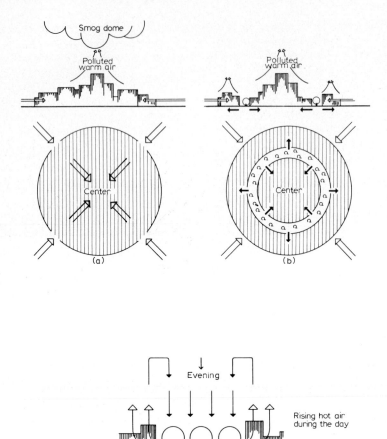

Fig. 56. Scheme of air flow in town at low aerial exchange weather conditions (calm days). (a) As a result of the rising heated air the field wind flows into the "low" formed in the town center, polluting it severely. (b) A green belt and other green areas clean and cool the air. Black arrows: fresh and clean air flowing from the green areas into the built up areas. (c) Air circulation.

greenery in the direction of the flow will purify and cool off the air, absorb carbon dioxide and release oxygen. The flow of warm, polluted air to the urban centre will thus be interrupted and the homogenous air stream of the field wind split up into several smaller circulations. As a result of thermal processes quantities of cooler air will repeatedly be emitted into the adjoining built up areas. The fresh air supply in these cases depends almost entirely on circulations which develop out of temperature differences between treeless and tree-stocked building areas. Technically, cooling aggregates of such sizes cannot be constructed (Figs. 56—58).

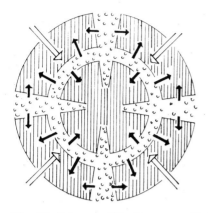

Fig. 57. Scheme of ideal town ventilation and green areas for cleaning and cooling the air. White arrows, increasing pollution and heating up; black arrows, supply of cleaned and cooled fresh air.

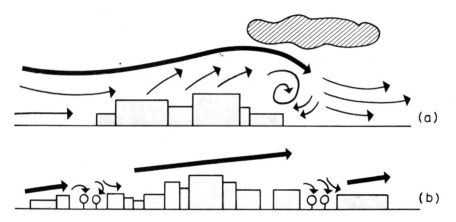

Fig. 58. Air flow in a town in windy weather. (a) Increasing pollution of the town from the main wind direction; (b) air cleaning by green areas.

5.2.7. Noise reduction by trees and shrubs

A special air pollution is noise. Some studies and investigations concerning the noise-reducting effect of plantation screens exist, but such investigations are particularly difficult because too many individual factors have to be observed. Much of what has been written on this subject is based on general considerations, without regard for the specific problems involved in selecting the right species of plants. Beck (1967) undertook to determine the characteristic noise-reducing capacity of different tree species. He also thoroughly discusses existing publications and comes to the following conclusions. The maximum noise reduction by plants amounts to about 10 dB, particularly in the frequency range of more than 1000 cps (=Hz) up to 11,200 cps (maximum at least at 8000 cps). Thus the effect of walls, embankments and similar installations is not equalled. The noise-reducing capacity of woods depends on the species used; in other words it is a specific characteristic. Beck lists different value-groups showing various degrees of noise reduction (Table 51). Decisive for noise reduction in the lay-out of screen plantings are the following aspects: plant leaves should be as big as possible and of a strong, hard structure; the leaves should overlap scale-wise and their position should preferably be perpendicular to the angle of incidence of the noise; foliage density is also necessary in the inner vegetation zone; deciduous trees which keep the dead foliage on their branches in Winter (hornbeam, oaks) are more effective than others which screen mainly in the summer months; evergreen conifers, though usually believed suitable "generally produce but little effect"; most of the broad-leaved evergreens suitable for the temperate zones produce a somewhat better effect due to their characteristic broad hard leaves, "whose specific effect is partly very good" (this applies especially to *Viburnum rhytidophyllum* and some *Rhododendron* species); plants for

TABLE 51

Noise reduction by trees and shrubs (after Beck, 1967)

Group III Noise reduction of 4 — 6 dB

Juniperus chinensis Pfitzeriana	*Lonicera maackii*
Betula pendula	*Crataegus* × *prunifolia*
Alnus incana	*Lonicera ledebourii*
Cornus sanguinea	*Acer negundo*
Cornus alba	*Populus canadensis* hybrids
Pterocarya fraxinifolia	*Corylus avellana*
Forsythia × *intermedia*	*Tilia cordata*
Sambucus nigra	

Group IV Noise reduction of 6 — 8 dB

Philadelphus pubescens	*Ilex aquifolium*
Carpinus betulus	*Ribes divaricatum*
Syringa vulgaris	*Quercus robur*
Fagus sylvatica	*Rhododendron*

Group V Noise reduction of 8 — 10 dB

Populus × *berolinensis*	*Viburnum rhytidophyllum*
Viburnum lantana	*Tilia platyphyllos*

Group VI Noise reduction of 10 — 12 dB

Acer pseudoplatanus

noise protection must have dense foliage from below upwards, particularly at the outer edge which is directed towards the source of the noise; shelter belts in successive parallel formation further noise reduction. About depth-efficacy, however, no concrete information exists. Cook and Haverbeke (1971, 1972) ascertained a noise reduction of 5—15 dB depending on the species of trees or shrubs, and on the height and width of the plantation. The effect may be increased by a land form or other solid barrier. Haupt (1973, 1974) studied the noise-screening effect of forests. Wendorff (1974) studied the problem of screening-off aircraft noise by forests and makes suggestions as to how this effect may be enhanced.

To sum up, shelter belts may serve as a valuable help in noise reduction provided they are designed systematically.

5.2.8. Oxygen release by trees in towns and forests

The oxygen reserve in the atmosphere amounts to 1200 billion (= American trillion) t (according to Roempp (1967) 1.18×10^{15} t; according to Johnson (1970) 1×10^{15} t). This is an enormous quantity compared with oxygen consumption. The common opinion is that this oxygen originated from the photosynthesis of green plants during the course of earth history. No exact data exist concerning the extent of photochemical separation of water vapour by ultraviolet light in the upper atmosphere, which was ascertained during the latest space flights. According to different authors the quantitative determination of the oxygen accumulated in the course of earth history is as follows: Johnson (1970) 13×10^{15} t; Welte (1970) 16.9×10^{15} t; Larcher (1973a) 17×10^{15} t.

Each year the oxygen supply of the atmosphere is replenished by a net production of about 70×10^9 t oxygen (= 70 billion t) from the assimilation of green plants (Larcher, 1973a). Statements by different authors about the yearly oxygen supply from the vegetation differ considerably (Table 52).

TABLE 52

Production of oxygen in billions of tons per year (after Bernatzky, 1973a)

Authors	Forests	Land plants	Seaweeds	Total
Bonner and Galston, 1952	28	45 = 10 %	360 = 90 %	405
Bruenig, 1971	55	—	—	—
Baumgartner, 1972	48	75 = 72 %	29 = 28 %	104
Cole, 1970				
Broeker, 1970	—	—	—	143
Peterson, 1970	1/3 of the O_2 production of all plants	—	—	—
Taylor,	—	30 %	70 %	—

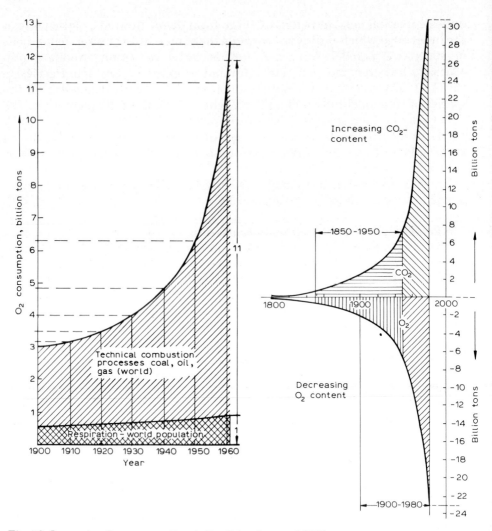

Fig. 59. Increasing O_2 consumption (after Schauberger, 1965).

Fig. 60. Technical combustion processes 1850/1950/1980, coal—oil—gas (world) (after Schauberger, 1965).

Up to the beginning of industrialization, the oxygen release by the vegetation compensated for its consumption by animals and humans, as well as for the decomposition of organic matter, in such a way as to leave a small surplus of oxygen, but a steady increase of oxygen consumption started with the burning of fossil fuels. Consequently the output of carbon dioxide rose as well (Schauberger, 1965; Straehler and Straehler, 1973; Figs. 59, 60; Table 53). The diminution of oxygen as a result of the combustion of all fossil fuels

TABLE 53

Increase of carbon dioxide concentration in the atmosphere since 1860 (after Anon., 1971, Umschau)

	1860	293 p.p.m.
	1950	306 p.p.m.
	1960	313 p.p.m.
	1970	321 p.p.m.
Estimated	1980	334 p.p.m.
	1990	353 p.p.m.
	2000	379 p.p.m.

is also estimated differently (Table 54). In part, the differences in this table can be explained by the fact that the available quantity of fossil fuels is estimated differently by different authors; indeed, new deposits continuously come to light. However, the whole oxygen and carbon dioxide problem has been investigated very little up to now. Exact measurements of the atmospheric amounts are not possible, so that all specifications are estimates only.

The consequences of the changes in the atmosphere are looked upon differently. Broeker (1970b) believes the oxygen reserve of the earth's atmosphere to be inexhaustible. He claims that if photosynthesis were ever to stop suddenly, it would take 10 million years to consume all the available oxygen, but Cole (1970) suggests 8000 years for this period. He does not think the oxygen reserves are unlimited, and points out that lack of oxygen may also be a local problem. He believes symptoms of oxygen crises to be possible in times of reduced atmospheric circulation. Baumgartner (1970, 1972) and Bruenig (1971) believe there exist immense oxygen reserves. Taylor (1970) quotes Johnson, according to whom the oxygen level of the atmosphere is very labile. Van Valen (1970) fears an irreversible loss of oxygen in consequence of the continuous destruction of the vegetation

TABLE 54

Diminution of oxygen as a result of complete combustion of all fossil fuels (after Bernatzky, 1973a)

Cole, 1970	8.0	%
Peterson, 1970a,b	0.6	%
Broeker, 1970b	few	%
Anonymous (Umschau), 1971	3	%
Baumgartner, 1972	<0.15	%

158

covering the planet, and stresses the irreversible nature of this situation. Peterson (1970a,b) refers to the worldwide land clearance and to the fact that the result of this process is a dangerous restriction of the binding of CO_2 by plants, so that worldwide alterations of climate have to be feared in consequence of increasing CO_2 concentrations. Recently Keller (1973a) published an oxygen balance for Switzerland. That country's yearly oxygen consumption (by plants, animals, humans, industry and traffic, as well as consumption by dissimilation) is estimated at 50 million t, or five times the oxygen release of 9.5 t. According to the author's careful calculation, highly industrialized Western Germany may consume at least ten times as much oxygen. But Anon (1970) calculated that the woods of Canada (ca. 250 million ha) are able to produce oxygen for 12 billion people; that is a quantity of 2.4 billion tons of oxygen.

Oxygen release by trees and woods

In view of oxygen diminution in the atmosphere, it is interesting to determine the actual oxygen release by trees and woods (Table 55). A 100-year-old beech tree, 25 m tall, with a crown diameter of 15 m (pages 21 and 24), a ground area of 160 m^2 and an external leaf surface area of 1600 m^2 (= sum of the individual leaf surface areas) produces 1.7 kg oxygen/hour while consuming 2.35 kg of carbon dioxide. The carbon fixed in its wood originates from the CO_2 of 40×10^6 m^3 of air (i.e. the content of 80,000 villas each having 500 m^3 enclosed space (Bernatzky, 1969b). On a sunny day, under optimum ecological conditions, a leaf surface area of 25 m^2 can release as much oxygen as a human being requires in the same period. However, since

TABLE 55

Oxygen release and carbon dioxide consumption of a 100-year-old beach (after Bernatzky, 1969b)

Height of the tree	25 m
Crown diameter	15 m
Volume of the crown	2700 m^3
Area underneath the crown	160 m^2
Outer leaf surface	1600 m^2
"Inner" leaf area (sum of the intercellular walls)	160,000 m^2
Volume of the wood	15 m^3
Dry substance of the wood (15 × 800 kg at 800 kh dw/m^2)	12,000 kg
Tied-up carbon	6000 kg
CO_2 intake	2352 g/h
H_2O intake	960 g/h
$C_6H_{12}O_6$ production	1600 g/h
O_2 release	1712 g/h

TABLE 56

NPP and oxygen release of various ecosystems (after Reichle, 1970 and Hansen, 1975)

Ecosystem	Area (10^6 km^2)	NPP of C (t/km^2/yr)	Total-NPP 10^9/yr	O$_2$ release (t/ha/yr)	Total O$_2$ 10^9 t/yr
Woodland or forest:					
Temperate "cold deciduous"	8	1000	8	26.0	20.8
Conifer "boreal a. mixed"	15	600	9	15.6	23.4
Rainforest "temperate"	1	1200	1.2	31.2	3.1
Rainforest "tropical and subtropical"	10	1500	15	39.0	39.0
Dry woodlands "various"	14	200	2.8	5.2	7.3
Subtotal	48		36	23.4	93.6
Nonforest:					
Agricultural	15	400	6	10.4	15.6
Grassland	26	300	15	7.8	20.3
Tundra-like	12	100	1.2	2.6	3.1
Other "desert"	32	100	3.2	2.6	8.3
Glaciers	15	—	—	—	—
Subtotal	100		18.2	5.8	47.3
Continents	148		54.2	14.6	140.9

160

photosynthesis does not occur at night nor in winter, at least 150 m² of leaf surface are necessary to meet the annual oxygen requirement of one person (Walter, 1950). Thus the beech tree mentioned above meets the annual oxygen requirement of ten people. If it is felled, 2700 young trees, each having a crown volume of 1 m³, would have to be planted to obtain the same result. That would require large sums of money (Bernatzky, 1969b). Usually, however, the ecological conditions are not ideal so that the oxygen surplus will not be so high. In any case, this calculation should prompt some reflection.

Hansen (1975) carefully analysed the results of the net primary production (NPP) and the assimilation of plants and trees as well as their oxygen release. On the basis of the respective NPP values, he calculated the corresponding oxygen amounts (Table 56). This table shows that according to the NPP results the oxygen release by woods is twice that of non-forest ecosystems. The conversion was done by Hansen. Accordingly, the total quantity of oxygen amounts to 140.9 billion tons/year, and that of the woods 93.6 billion tons/year. These data correspond well with those calculated by Whittaker (1970) (138.2 and 101.9 billion tons/year, respectively). On the basis of the NPP calculations by Lieth (1972), adopted also by Larcher (1973a), Hansen arrived at the final figure of 140.4 billion tons/year representing the total amount of oxygen produced by all land plants on earth. According to the above authors, forests therefore produce the quantities of oxygen given in Table 57. According to this table forests produce 2/3 of the oxygen of all land plants. The value calculated by Bruenig seems too low. In the International Biological Program (IBP) the most complete report so far about the production capacity of a beech tree (*Fagus sylvatica*) was published by Schulze (1970) as part of the Solling Project. Using Hansen's conversion of the NPP values, it shows a total output of 16.63 t O_2/ha/year. For details about the methods and results of ecosystem research in the German Solling Project see Ellenberg (Ed.) 1971. The amount measured by Schulze corresponds with Schmid (1972), who gives the oxygen release of coniferous forests as 30 t/ha/year and that of deciduous forests as 16 t/ha/year (as against agricultural areas: 3—10 t/ha/year). Table 58 shows the forest's production as calculated by Polster (1950). When converted to leaf surface area

TABLE 57

Oxygen release of the forests on earth (from Hansen, 1975)

Author	10^9 t O_2/year	In % of all land-plants
Bruenig, 1971	55	—
Lieth, 1972	85	60
Reichle, 1970	94	67
Whittaker, 1970	102	74

TABLE 58

Production of the forests of the temperate zone (after Polster, 1950 and converted by Hansen, 1975)

Species of tree	Gross production t C/ha/yr	Net production t C/ha/yr	Net production in % of gross production	Oxygen release t O$_2$/ha/yr
Birch	13.5	3.6	27	9.5
Beech	17.1	5.9	34.5	15.6
Oak	10.2	4.3	42	11.5
Larch	15.1	5.9	39	15.6
Douglas fir	30.8	7.3	24	19.3
Spruce	10.0	5.9	31	15.6
Pine	8.7	3.9	45	10.3

the values change, as shown in Table 59. Hence deciduous trees — that is, the leaf surface areas — release the largest quantities of oxygen. Bonnemann and Röhrig (1971) calculate the oxygen production of the temperate zone as follows: deciduous forests 6—10 t/ha/year; coniferous forests 9—15 t/ha/year. For tropical and subtropical forests the average values are given in Table 60. Despite the far higher substance production in the tropical rainforests, a net oxygen surplus of 18 t/ha/year corresponds with that of a 100-year-old beech wood in Central Europe; the responsible factor being the significantly higher respiration rate (Walter, 1970). It is to be borne in mind as a matter of principle that all specifications given are only approximate values from which, depending on prevailing ecological conditions, deviations are possible.

Now evidently the oxygen release by trees and forests is a minor affair compared to the large reserves in the atmosphere. We hear and read again and again that the oxygen loss in the biosphere, which results from domestic fuel combustion and from traffic and industrial activities, could be sufficiently compensated by exchanges in the atmosphere, but in fact this is doubtful. In quiet weather conditions (wind speed below 2—3 m/sec, inversion, etc.) there is little or no air exchange. The supply of oxygen is

TABLE 59

Oxygen release converted upon leaf surface (after Hansen, 1975)

Birch (*Betula pendula*)	2.53 g O$_2$/dm^2/h
Beech (*Fagus sylvatica*)	1.76
Oak (*Quercus robur*)	1.53
Larch (*Larix europaea*)	1.70
Douglas fir (*Pseudotsuga douglasii*)	1.12
Pine (*Pinus sylvestris*)	1.20
Norway spruce (*Picea abies*)	1.15

TABLE 60

Oxygen release of tropical and subtropical forests (after Hansen, 1975)

Author	t O_2/ha/yr
Weck, 1959	25
Lieth, 1964/65	27
Rodin and Bazilevich, 1967	23
Reichle, 1970	35
Whittaker, 1970	26
Bruenig, 1971	29
Lieth, 1972	26
Walter, 1973	29

then severly limited or stopped altogether. Under these conditions, too, more oxygen is burned and more CO_2 released into the air. Moreover, the increasing quantity of poisonous substances in the air shows that the cited air exchanges are not sufficient. For that reason local oxygen deficiencies must be feared (Cole, 1970), which may seriously endanger the health of people living there. Taylor (1970) mentions a deficiency of 6% in Los Angeles. Unfortunately, exact oxygen measurements of the living space of city dwellers, that is of the air space up to 100 m above the ground, are as yet unknown. It is high time that urban ecology tackled this problem. Russian scientists are intensely occupied with the O_2—CO_2 problem. Davitaya (1971) thinks that if oxygen consumption stays on the same level, or if the

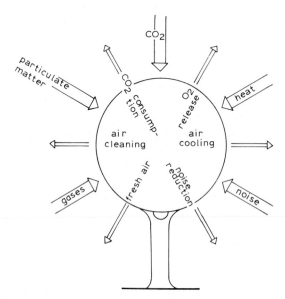

Fig. 61. Scheme of tree functions.

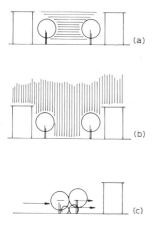

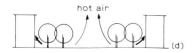

Fig. 62. Functions of street trees. Street trees (a) prevent particulate matter from entering flats and offices; (b) absorb, in the evening, the sinking particulate matter coming down with cooling off; (c) filter noise and pollution coming from the street; (d) cool the hot air in the street.

yearly decrease amounts to 10 billion tons, then after 100,000 years 2/3 of all the free oxygen in the atmosphere and the hydrosphere will be exhausted. By then the amount of CO_2 in the atmosphere will have reached dangerous concentrations. In fact, however, the consumption of oxygen and the concentration of CO_2 will increase further. With a yearly increase of 1%, the critical situation will arise in only 7000 years, with a 5% increase it will take 180 years, and with 10%, only 100 years. As mentioned before, the increasing CO_2 concentrations in the air, together with the oxygen loss, cause the danger of rising temperatures all over the world (glasshouse effect of atmospheric CO_2). This again shows how intricately interwoven ecological factors are. "Insignificant operations at one place may have unexpected and serious ecological effects in an altogether different place. We utilize more than 40% of the total land area and have reduced the total amount of organic substance of the land vegetation by one-third. These quantities are perfectly capable of inducing a change in our ecological system" (Anon., 1971). Fifty years ago everybody believed that water was available in unlimited quantity and quality. Today we have an analogous problem with the air and its oxygen. Therefore every single tree in town and its surroundings helps to prevent local oxygen deficits (Figs. 61 and 62).

164

5.3. Effects of trees on physical and mental health

People generally know that green areas and forests further good health. Indeed they regard them almost as a symbol for health, but generally speaking, clear notions as to exactly how trees and health are related do not exist. The beneficial effects of trees are overestimated by some and underrated by others. The difficulties start with the definition of health. Freedom from illness is a negative definition and moreover unclear. There is an indefinite number of illnesses. WHO defined health as not only the absence of illness, but also as the condition of absolute physical, mental and social well-being,

Plate 37. The best recreation for man is under trees! (Photo Stummer).

Plate 38. Healthy children in beautiful parks (Photo Bernatzky).

but that is to be regarded as an aim to strive for rather than as a definition. Nonetheless, this definition is very helpful for our purpose. Because trees, green areas and forests are especially beneficial in this respect, whoever is sound in body and mind can also work efficiently (Plates 37 and 38).

The climate effects of trees were discussed on page 145 onwards. It is now important to consider these climatic functions in relation to their effects on health. The best way to do this is to point out that the negative results of climatic deterioration are reversed by the effects of trees. The long-term influences of climatic conditions on human health are investigated by bioclimatology. The applied form is bioclimatics. Short-term weather conditions and their effects on biological events are investigated by biometeorology (Weickann and Ungeheuer, 1952; Seybold and Woltereck, 1952). Observations that such atmospheric influences on the progress of illness exist, are very old. Systematic investigations, though, are relatively new. One difficulty, among others, is the fact that the vital processes of a single individual are so closely interrelated that it is almost impossible to trace a single one by itself. Thus these interrelationships still require thorough investigation. Even so, many incomplete findings do exist so that we are no longer entirely in the dark (Martini, 1952; Pfleiderer and Schittenhelm, 1952; de Rudder, 1952; Vogt and

Amelung, 1952; and others). In what follows the interrelations between individual climatic factors and health will briefly be outlined.

The sun's radiation

Visible light takes effect mainly in the psychological domain. Radiating sunshine stimulates and invigorates. The territories in the North of the globe with their bright summer nights and their pitch-dark winter days show this very clearly, and also the opposite reaction caused by lack of light. Tests have shown that ocular reception of light affects the hypophysis, thus causing alterations in the production of hormones. Interrelations also exist between the nucleus praeopticus and the hypophysis. Moreover, processes are triggered off in the visual purple of the eye, leading to the development of vitamin A (de Rudder, 1952). In the city the soothing colours green and blue are absent, whereas the proportion of the exciting light energies red and yellow are increased. In connection with the particular influence of single light-wave ranges, very unfavourable constellations develop (Hellpach, 1965). Daylight in the city combines heightened, psychologically exciting energy with reduced physiological activity, whereas the light in woods, green areas and the open country, by contrast, spurs physiological activity but is psychologically soothing. The city dweller lives increasingly with artificial, super-glaring light.

Ultraviolet light has a particular effectiveness which is, for the most part, lost to the city dweller. It has been proved that ultraviolet radiation plays an important role in preventing rickets. Furthermore, it promotes the multiplication of free sulphohydric groups (Sulfhydrilgruppen), which are activators of many enzymes and ferments as well as of important hormones (Amelung in Vogt and Amelung, 1952). Under intense UV-radiation the thyroid gland works differently than when there is UV deficiency. The trees and green areas of a town lie beneath the smog dome developing above it which, as we have seen, reduces UV-radiation. This reduction also affects all trees and green areas in the city, but the green parks and gardens stimulate the population to leave their homes and stay in the open air where they will, after all, receive a larger dose of UV-radiation than those staying in the deep UV-shade of streets and buildings, or behind the walls and windows of their homes. It is a very particular value of trees and other green places that they invite people to go outside and stay in the open air. Who knows whether modern city dwellers spend even 10% of their lifetime in the open air? Perhaps this figure is even lower, considering the number of motorized vehicles.

Temperature

It is well known that overheating (hyperthermy) causes disturbances of health: congestion of blood to the head, headache, nausea, fatigue, exhaustion to the point of unconsciousness. The high mortality rate of babies in the summer is due to this overheating (de Rudder, 1952). The city lacks cooling facilities. A gentle climate develops, with hardly any stimulation. This, together with air-conditioning, weakens the human body so that it easily falls victim to local cold spells (increases of "colds" and influenza).

Air humidity

The higher air temperature corresponds with lower relative air humidity, especially in summer. This leads to an increase of illnesses of the nose and the upper respiratory organs. Relative air humidity might be raised in the summer when green areas are watered, without the limit of mugginess being reached (Bernatzky, 1960).

Diminution of oxygen

Normally disturbances of the vital functions of the human body occur only at an 80% oxygen saturation of the haemoglobin. This almost corresponds with oxygen conditions at an altitude of 4500 m. At a height of 6000—7000 m the overall conditions become dangerous. Unfortunately, comprehensive measurements of local and intermittently terminated oxygen deficits have not been made in our cities. (For the entire problem see page 155 onwards.) It is high time to start them. In this connection it must be pointed out that damage is seldom caused by a single factor but rather by several factors together. Therefore an oxygen deficiency may be of importance, particularly with inversional weather conditions entailing interrupted air exchange with higher layers of the atmosphere.

Air pollution

Here matters are now clearer, though many questions remain to be answered. To begin with, particulate pollution serves as nuclei for the condensation of water and poisonous gases of the air. Condensation of water vapour on these particles of aerosol adds to the development of town fog. Breathing depth is diminished through this and psychological disposition is adversely affected. The smog catastrophes in the Meuse valley (1930), Donora, Pennsylvania (1948), Los Angeles (1947 and 1949), Poza Rica, Mexico (1950) and in London (1950 and 1952) were caused by a severe increase of the SO_2 content of the air due to dusts and smoke. In the 1952 London smog, 4000 more people died than were statistically expected to do so (Pott, 1972). Analogous climatic conditions in the Ruhr District of Germany produced similar results in 1962. Patients already weakened by diseases of the respiratory system and the heart, as well as those with circulatory trouble, fell victim to the air pollution. Zweimann et al. (1972) state that SO_2 and particulate material were associated with increased morbidity and mortality from chronic obstructive bronchio-pulmonary disease, especially in older patients. A high level of nitrogen oxides has also been associated with increased pulmonary infections in young children. The toxicity of SO_2 is the result of its transformation into sulphuric acid. It irritates the skin and mucous membranes, particularly of the conjunctiva and the bronchial system, and causes coughing and vomiting. Bronchitis, chronic changes in the lungs and even a lung oedema may result (Becker, 1971). Investigations by Kapalin (in Becker, 1971) in Czechoslovakia proved that the life expectancy of children in industrial regions is reduced by five years compared to that of children growing up in clean air. Maturing of bones, and with it overall growth, was clearly delayed in comparison with children growing up in the country.

Substances isolated from soot, from automobile exhaust fumes and from cigarette smoke have been found to lead to the development of tumours in animals, but possibly humans are more sensitive than experimental animals and possibly combinations of several substances are decisive for the toxic effect; also the accumulation of different substances plays a part. For instance the carcinogenic effect of benzopyrene may be increased by adding other, non-carcinogenic, substances. Benzopyrene stays significantly longer in the lungs if attached to dust. On the other hand, it has been observed that the number of tumours caused by a benzopyrene—pollen mixture was smaller than that caused by pure benzopyrene (Pott, 1972). The connections between air pollution and lung-cancer were investigated by Hettche (1970). The carbon monoxide in the exhaust of motor vehicles possesses 300 times as much affinity to the haemoglobin of the blood as does oxygen; that is, it adheres to and accumulates in the haemoglobin 300 times sooner and faster than oxygen. CO causes visual disturbances, psychological indisposition, apathy and headaches leading sometimes to the nap behind the wheel (limousine-sickness). A prolonged stay in air rich in CO causes irreversible disturbances such as disorders of the respiratory system and of the central nervous system, muscle convulsions and decrease of blood pressure to the point where comatose conditions arise. Also, small permanent concentrations of 20% CO-Hb further the development of arteriosclerosis. Nitrogen oxides cause irritations of the mucous membrane and burning of the eyes. Higher doses may give rise to pneumonia and oedema of the lungs. Fluorine pollution of the air may affect the joints, spine, muscles and nerves and cause intestinal disturbances.

Lead

Well-known are sicknesses caused by lead which gets into the air through the use of lead-containing petrol for motorcars (Stoefen, 1974). The world-wide, yearly quantity of lead sent into the air by the combustion of power fuel is estimated at more than 500 million tons. The lead in the air is taken in by plants, animals and humans. It is stored in the bones, where it may, under given circumstances, be mobilized and enter the blood. It then impedes enzymic systems which are necessary for vital processes and for the synthesis of blood pigment. A continuous supply of lead will give rise to a lingering toxicosis. Symptoms of poisoning are headaches, insomnia, bowel troubles and even colics, constipation, loss of appetite, rheumatic complaints, muscular weakness, metabolic diseases such as lead-basedow and anaemia.

Other investigations concerning the effects of air pollution on health have been made by Mellanby (1972), Neuwirth (1974), Zweimann et al. (1972).

Noise

Noise at grade I, of 30—60 dB, mainly causes psychological stress. It is enough to severely disturb sleep. Grade II, of 60—90 dB, endangers health by psychological and vegetative effects. At grade III, of 90—120 dB, noise damages health not only by psychological and vegetative reactions but also by those of the organ of hearing. Above 120 dB, at grade IV, lies the threshold of pain for the ear (Becker, 1974). The ranges mentioned merge into one another.

Summary. In as much as trees and other green areas contribute to the diminution of the types of air pollution mentioned (page 135 onwards), they render human health a valuable service.

Change of the environment in city life

Apart from climatic factors and air pollution, the total psycho-physical condition of the city dweller has been changed. The city detaches its inhabitants from the country and from nature; in city life one is spatially removed from nature. Cities have expanded so much that a stay in open country has become more and more difficult. Immediate contact with nature is difficult. Mental alienation from nature follows in the wake of statial alienation. The natural rhythm of vital processes is disturbed in the city and it is an important general base of our existence. Amelung (1952) states that "the conservation of the natural rhythm is one of the most important requirements for a healthy life. The uninterrupted rush of modern civilisation, the irregular intake of food, the change in normal sleeping time, the illumination of the city and the blotting out of seasonal differences in terms of food and way of life, endanger the natural rhythms of humanity". The screening of the soil against electricity from the earth and atmosphere and against natural radioactivity is of great importance to human health, according to Hellpach (1965). Psychological and physical lability are especially at home in the towns. Piperek (1957) found that in the highly technological city centres of North America, England and Sweden, the number of registered cases of neurosis, depression, irritability and weakness, and the incidence of cases where feelings and a sense of values have atrophied, is especially high where natural influences are almost completely removed.

Nature — progress or retrogression?

Trees and the green countryside are basic forms of human experience. However, our technological and artificial environment blinds us towards nature and many of us believe we can do without it. Plastic trees are thought to suffice as a phantasmagoria of nature! (Krieger, 1973). Evidently, however, technical progress entails a considerable narrowing of the biological foundations of human life. If humans and animals are taken out of their natural habitat where they still relate to natural environmental conditions, their development is distorted. Lorenz (1973) has shown that a biological shift of reference then occurs. This would be followed by catastrophic results if no human and animal adaptations could be developed by means of numerous adjustments. However, human adaptability is limited and, in the opinion of some physicians and scientists, already surpassed. Moreover, such acquired characteristics cannot be inherited so that each new generation is forced to acquire them again — for sheer survival!

There is, therefore, no other way but to submit to the laws of human nature. Many call this a return to natural human fundamentals. Is this notion illusion or necessity, recession or progress? The psychosomatic condition of humanity answers this question.

The first stimuli for basic forms of psychological human experience go back to impressions of surrounding nature, with which humans had to deal on their way through hundreds of thousands of years. The human psychological condition is most deeply rooted in the sphere of experience and impressions of nature. The technological age signifies only the last minute of this, our first day of life, and it is obvious that the abrupt change from a natural environment to an artificial milieu cannot but give rise to psychological problems of adaptation. All of our essential psychological functions were originally awakened and trained by the resonance of cosmic impulses in the human psyche and by the interaction between mind and nature (Piperek, 1957). According to the theory of psychic resonance, the objects of our surroundings act like a force field. Piperek investigated this and found that the mental effect of experience deriving from natural environment was positive in 55—85% of all cases. In contrast, experience grounded in the city milieu had a negative effect on 55—62% of persons tested. As natural objects of experience must make way for technical objects, the experience of resonance likewise becomes technical. Accordingly, human experience comes increasingly to consist of a mass of one-sidedly technical impressions to whose number and intensity humans cannot adapt biologically without injury to themselves.

Trees and forests, however, are a means of return to our biological origin. A stay under trees and in green surroundings always means a change of scene, so often prescribed by physicians for human relaxation. However brief such a stay may be, it remains a confrontation with "the wholly other", that is nature. In the presence of trees, city dwellers again take part in the rhythm of day and night so that their minds come to rest and they find peace. We know from psychology that a certain impression, or mental experience, can profoundly affect all or many subsequent impressions or even make them glow in a positive or a negative sense. The mental experience of the beauty and harmony of nature likewise carries over into other experiences and daily tasks. According to Mitscherlich (1969, 1971) humans do not lose their bond with primeval nature even at the heart of technological civilisation. Remnants of nature in towns are the trees, parks and other green areas. They are the absolute minimum requirements if a miserable form of human existence is to be avoided. A treeless town will, in due course, become an unbearable strain for those living in it. No wonder that in the many and oft-repeated inquiries, the majority of city dwellers vote for green areas in town and in the immediate vicinity.

6. GENERAL PHYTOPATHOLOGY

A tree at a location which is, in a comprehensive sense, the right one (sum of ecological factors) and which is damaged neither by natural forces nor by man, directly or indirectly, will theoretically always remain free from disease, or be able to fend it off. However, this theoretical case is hardly ever real. Cultivated plants and trees, especially, suffer from a disturbed biocenotic balance. The alteration of nature by man leads to disturbances of physiological processes (diseases) and with them to an intensified susceptibility on the part of trees to all kinds of pests; nature thus affected can therefore be kept in existence only by a continuous struggle against the environment (Kirchner, 1967). However, the number of pests and diseases is on the increase, especially in monocultures.

In natural plant communities or in the natural distribution area of a tree species resistance against prevalent diseases has attained an equilibrium during the long course of phylogenetic development. The predatory and parasitic enemies of the insect pests attack the latter at once until an equilibrium is achieved. Outside the natural distribution area this equilibrium gets lost. The tree becomes diseased. This happens when, for instance, conifers are planted in soil which would naturally grow deciduous trees (Schwerdtfeger, 1970). On the other hand, the infestation of trees with pests and diseases in the unsuitable conditions of an artificial forest or a town location is the self-adjusting answer of the ecosystem, which abandons the species that are unsuitable and so aids their destruction. The forest or tree as a relational network is composed of an immense number of organisms and cannot be disturbed by a small number of pests. Only a mass of them produces problems.

Abiotic causes of diseases

Climate, weather, light, shade, heat, frost, natural forces, rain, snow, surplus or lack of water and lack of nutrients are some of the most important abiotic causes. They may be warded off by improving the location of trees. The less favourable their location, the more will trees also be subject to biotic causes of diseases.

Biotic causes of diseases

These include viruses, rickettsiae, bacteria, fungi, lichens, microsporidiae and arthropods as well as insects, birds and mammals. Viruses cause the mosaic disease on beech, birch, ash, maple, elm, rowan-tree, poplar and locust. Bacteria and rickettsiae play a subordinate role in causing tree diseases; mostly they act as wound parasites and enter the plant via external injuries. They live in the intercellular spaces and in the vessels which they clog. Only a few of the bacteria enter the living cells. The infected plants often develop hypertrophical growth (e.g. bacterial cancer on poplars). Viruses and bacteria are in many cases antagonists of noxious insects. Particularly to be mentioned is *Bacillus thuringiensis* (page 311), which is used in biological pest control against many lepidopterous insects. Together with insects, fungi are the most serious causes of tree diseases. They can also infect insects. It is a fungus, too, which causes the dreaded Dutch elm disease (DED). Its name is *Ceratocystis ulmi*, a member of the *Melanosporaceae*. This disease may kill trees within months, or over the course of several years. The elm splint bettle in its burrows infests the tree with the fungus, which in turn blocks the vessels and stops water transport. Particularly elms in unnatural locations, such as parks and streets, fall victim to this fungus. The agents of cancer (*Hypocreaceae*) also belong to the fungus family. The mycelium of this fungus kills the external tissues and stimulates the adjoining ones to excessive growth. The resulting large bulges create the "cancerous ulcer".

Fungi are capable of producing incredible numbers of spores. Some species produce no fewer than 2500 million spores hourly (*Fomes tomentarius*). Fortunately the endangered tree must meet certain requirements, just as on the side of the infecting agent a certain disposition must be there to start an infection. Spores are produced within a temperature range of −5° to +40° centigrade. The optimum lies between 20 and 30°C. The period of fructification, too, is temperature-controlled. High degrees of humidity in wood and air favour infection by fungi; light generally inhibits it. The formation of the mushroom body of the fungus always takes place in the light. The growth of the fungi depends on nutritive content and on the pH-value of the substratum. Germination of the spores dispersed by water, wind, animals and humans, requires relatively high air humidity (dew, fog). Conditions favouring tree infection by fungi always exist. Lichens, those hybrids produced by fungus and alga, can indirectly damage trees by furthering the rotting of the bark and so on. Dicotyledonous parasites such as *Viscum album* and *Cuscuta* can kill trees.

The second large group of pests on trees comprises widely varying kinds of insects which chew, lick, sting or suck; they are particularly dangerous in the grub stage. In this stage the caterpillars of the pine moth can multiply their starting weight about 900 times. While mass infections by fungi may occur at any time, those of insects need a running-in period unless immigrations from other areas take place or local climatic and environmental conditions favour their rapid mass appearance.

Birds are generally useful to trees because they fight noxious insects.

Schwerdtfeger, however, wonders whether this usefulness is not cancelled out by the reduced multiplication of specific parasites which control the insect pests and then can find no more food. Finches, moor hens and wood peckers can cause damage by biting buds and peeling bark. Mammals, especially rodents and voles, cause damage to bark and roots.

Disposition of trees towards diseases

Disposition towards diseases depends not only on the characteristics of a tree species, but also on given circumstances, on the nutritive condition, on existing damage, on the stage of development and on the season. Alterations of the location by man, such as a change in the ground-water level or airtightness in the soil, increases the disposition towards diseases. Such factors reduce the vitality of the tree and so diminish pest resistance. Then only an injury is needed, enabling pathogenic organisms to enter the tree, for an infection to occur.

Resistance

There is passive as well as active resistance; each merges into the other. Passive resistance consists of mechanically protective characteristics such as a tough cuticle, pricks and thorns (which protect against animals), thick bark, the presence of certain substances (repulsive, biostatic or biocidal), and resin. Active resistance includes processes that are triggered off by an attack of the infecting agent. Examples are overgrowing of the diseased spot, the formation of new tissues obstructing the entrance of fungus mycelium, and the production of wound-cork and wound-wood.

Disposition towards diseases is intensified in monocultures where many trees of the same species are neighbours. Stocks of mixed species show better resistance. On the other hand, a mixture of species makes it possible for some infecting agents to change hosts. Much damage appears in certain age groups so that stocks comprising various ages are more resistant. Dense stocks further fungus infections because of higher relative air humidity. As soon as the biocenosis in a stock of trees is disturbed, the equilibrium is lost and pests can multiply far more easily and quickly. The result of an attack by infecting agents is a physiological alteration in plant metabolism: increased or reduced transpiration, respiration, or assimilation; disturbance of the water and nutrition balances. Toxins are spread by the infecting agents. These effects alter the material composition of the host (the carbohydrate and albumin content; cell sap acidity; resins, gums and other excreta are formed). The normal life rhythm of the plant is profoundly affected.

7. TREE PRESERVATION

"Where building takes place, trees die!" Building activity in town and country (roads as well as buildings), has proved to be the greatest enemy of trees. In most cases, it is not so much a question of overt hostility towards trees (although this does exist), as of ignorance of the living conditions essential for trees, and of the effects of trees on the maintenance of a tolerable environment. In many cases protective measures for trees are well-known but not carried out because time and money are involved. In this way countless trees die or are removed which might have been saved by reasonable planning and consideration of the ecological necessities. Trees are not just dead wood, they are living beings. They are just as sensitive as people and animals to disturbances in their way of life.

7.1. Why trees die (Plates 39—48)

There are a number of constantly recurring causes of damage which can and do lead to loss of trees (Figs. 63 and 64).

Lowering of ground water. In building operations the ground water is frequently exposed, so to enable work to continue it is pumped away and the water table is considerably lowered. The tree roots, which are adapted according to species and habitat to a specific level of ground water, are then drained. The trees die from lack of water.

Soil compaction. Just as great are the perils of soil compaction in the root zone of trees. Site huts are set up under trees in most cases. Then for months workers tramp over the root zone of the trees, construction vehicles with their heavy loads of materials compress the soil, and earth moving equipment pushes the soil away thus destroying roots. Excavated soil is piled up and spread out over the root zone. As a result of all this, the cavities in the soil are pressed together and thus ventilation of the soil is impeded or made impossible. Compaction deprives the millions and billions of minute organisms in the soil, without which a tree cannot live and find nourishment, of air and living space and the trees die from lack of soil oxygen.

Removal of topsoil. The soil is not just ground for building on; not just an inert mass. On the contrary it is a huge dynamic unit in which physical,

Plates 39—44. How can trees survive? (Photos Maurer).

chemical and biological processes go on incessantly. Since the tiniest organisms are found in the uppermost centimetres of the soil, removal of topsoil in the root zone has an exceptionally harmful effect; nutrients and their processors, the bacteria, are lost and the trees die from lack of nutrients.

Root losses. Damage to a tree is even greater if the roots themselves are attacked and destroyed due to excavation. When drains and cable trenches are laid, or when road strengthening, fencing and foundations are constructed, many roots are destroyed; this applies to strong supporting roots

Plate 45. Trunk damage caused by vehicle (Photo Bernatzky).

which are indispensable for vertical stability, as well as to the fine fibrous roots which feed the tree. Excavators digging a hole, tear off roots deep down in the ground, or crack them so that the damaged parts cannot even be cured. The root areas of the trees become covered with tar, asphalt and concrete. In this way the roots are denied air, water and nutrients. Road-strengthening works using cinder cause greater damage due to the leaching out of sulphur. Pollution of the soil by oil, corrosive fluids (flushing water from concrete mixers and so on) poison the soil and its organisms. De-icing salt in Winter and illuminating gas (natural gas) are especially dangerous.

Damage to trunk and leaf crown. Damage to the trunk occurs where stones are piled against it, bruising the bark; vehicles rub off large pieces of bark, and fencing and cables are nailed to trees. The tree crown suffers damage due to the burning of refuse and to the erection of site huts having solid fuel stoves. Isolation of trees from a group leads to sunburn (destruction of the cambium) and to wind breakage.

Trees are poisoned by
de-icing salt
town gas and natural gas
oil

Plate 46. Root damage caused by excavator (Photo Bernatzky).

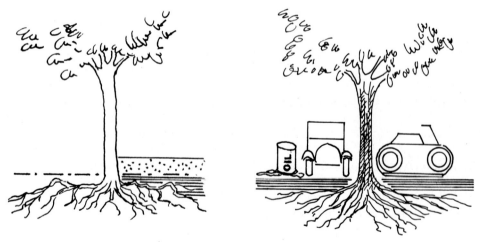

Fig. 63. Tree damage by soil removal of overfilling.

Fig. 64. Tree damage by compaction by vehicles, oil, chemicals, etc.

Plates 47 and 48. Suffocating trees (free root area too small) (Photos Bernatzky).

chemicals
immissions of noxious gases (SO_2, nitrogen oxides, HF and others).
    Trees are mistreated by
root damage in the course of construction work

180

trunk damage by cars
windbreak
false pruning.
   Trees do not tolerate
overfilling with soil
excavations
concrete and asphalt coverings of the root area
grass cover
lighting of fires under the tree crown
isolation leading to sunburn
lightning
frost cracks in spring.
(Literature for tree surgery: Fenska, 1964; Pirone, 1972; Bridgemann, 1976.)

7.2. Standards of tree maintenance

7.2.1. Pruning of trees (Plates 49—54)
   The healthier a tree, the fuller its crown and vice versa. We have seen that the crown with its leaves is the workshop of the tree and the source of all beneficial effects to man. Regular pruning of a tree crown, therefore, is always wrong. It weakens the vitality of the tree as a whole because an essential part of its productive faculties is taken away (Fig. 65). The more frequent and severe the pruning, the more wounds occur, giving access to thousands of fungi and bacteria. As little pruning as possible will keep the tree healthy. However, other rules apply to short-lived fruit trees.

7.2.1.1. When to prune? Actually there are only a few reasons for pruning a tree:
(1) Pruning to aid recovery from injury.
(2) Pruning to reduce the crown load after root losses and tree surgery.
(3) Safety pruning of old trees (for reasons of stability, but only very rarely) and of tree crowns projecting into the space of main roads which have to be kept clear (usually 4.50 m clear height) and in the case of overhead wires.
(4) Pruning to thin out wherever there is too much shade and too close planting.
(5) Special pruning in cases of tree surgery.
(6) In addition, and as a matter of principle, all dead or dying branches, shoots impeding growth (water shoots, stool sprouts) and branches crossing and rubbing each other should be removed.
   Each conventionally trimmed and lopped tree crown, i.e. a crown completely robbed of its branches or with its branches severely cut back, will produce new growth that it far more bushy and dense than ever before. At one cut five or more branches develop instead of the one that was there

181

Plates 49—54. Obsolute pruning. 52 and 53, The same tree as Plate 51; the crown is growing more dense. 54, "Clothes-hangers" (Photos Bernatzky).

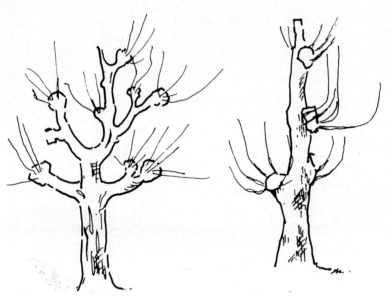

Fig. 65. Trees stunted by incorrect pruning.

before. That is why far more shade is produced by the new crown. New branches break off easily at first, and the big, mostly untreated, cuts represent entrance gates for the vast army of fungi and bacterias. If large sections of branches have to be removed to reduce shade or to make up for root losses or for similar reasons, then the thinning type of pruning should be carried out. It is better and cheaper than the usual mutilation of trees. The method consists of cutting a twig or branch right at its base at the nearest stronger part (Fig. 66). Wounds then heal better and faster because they get fed better and the usual whorls are absent.

Pruning should be done only by someone who has the necessary botanical, physiological, ecological and biological knowledge for this work. Naturally the characteristic peculiarities of the trees to be pruned have to be considered. This means that bleeding trees (such as maple, birch, dogwood, walnut and elms with yellow wood) should be pruned only at times of low sap pressure, i.e. late in Summer or Autumn but not in Spring. Conifers generally do not sprout again at cuts (pines, spruces, firs). If with these trees the leader is broken off or cut away by mistake, a lateral branch can be tied vertically to a stick and kept in this position for about two years. It will then develop into the new leader. Evergreens of the *Arbor vitae* class (junipers, hemlock, cypress, false cypress, cryptomeria, larch, golden larch, *Podocarpus*, *Thuja*, *Torreya*, incense cedar) may be pruned at any time. Flowering trees blossoming on annual sprouts can be pruned in Winter or Spring, but those flowering on two-year-old shoots should not be pruned in Winter and Spring because

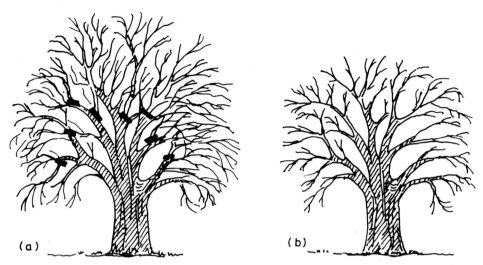

Fig. 66. Thinning type of pruning. (a) Branches and twigs are cut off right at their base; (b) result of pruning. The tree crown has retained its shape, but has become thinner and lower.

the buds already existing would then be cut off. They should be pruned immediately after flowering. Palms should never lose their terminal bud and care should be taken not to remove it, since otherwise the tree will die (see also Brown, 1972).

7.2.1.2. How to prune? Pruning is done in stages. About 30 cm beyond the intended cut, the branch to be removed is sawn into from below until the saw begins to stick. Then the next cut is made close to the first but from above and still beyond the intended place. One single cut from above would cause the weight of the branch to tear big pieces of wood and bark from the tree (Fig. 67). The stump remaining after the second cut is now removed from beneath, as closely as possible to the trunk or main branch, on sap flow level. Because it is not always possible to get close enough with the saw, one has to use a chisel afterwards. The wound edges, which are always roughened by the saw (destroying whole cell systems), should be cut smooth with a sharp pruning knife. A cut at right angles to the longitudinal axis of a branch would certainly reduce the cut area, but healing is poor because the supply of assimilates for creating the callus, which is conveyed in the downward sap flow of the phloem cannot flow up a stump. As a result callus does not form there, the wood dries up, water and fungi invade it and cause rot which may penetrate deep into the trunk.

Cut surfaces of over 5 cm in diameter have to be enlarged upwards and downwards to a pointed oval shape (elliptical) to facilitate the encircling flow of sap which supplies the nutrients for healing the wound (Plates 55—

184

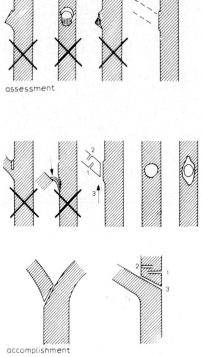

assessment

accomplishment

Fig. 67. Assessment and accomplishment of pruning.

58). The callus always grows along the line of the flow of nutrients. Cuts leaving behind socalled "clothes hangers" (ugly stumps) should be avoided. They soon start to decay and give access to a lot of diseases as already mentioned.

7.2.1.3. *Wound healing*. The killing of cells causes the nuclei of neighbouring sound cells to move to the cell walls next to the wound. This increases respiration and the body temperature of the wounded tissue is raised by one to three degrees centigrade (Huber, 1962). If young tissue is injured, then the repeated parting of cells gives rise to a proliferation called callus in the layers adjacent to the wound. The less differentiated cells of the parenchyma in particular may go into division again in this process. In normal trees the healthy part of the cambium adjoining the wound produces the callus. After its formation, differentiation begins again in this proliferation, meaning that everything lost is going to be restored (vessels, tubes, etc.). The tissue bulge of the callus closes up outwards with cork; inside, a cambium layer is formed which connects with the existing cambium of the trunk and functions like it. Furthermore, the wound is covered with new layers of

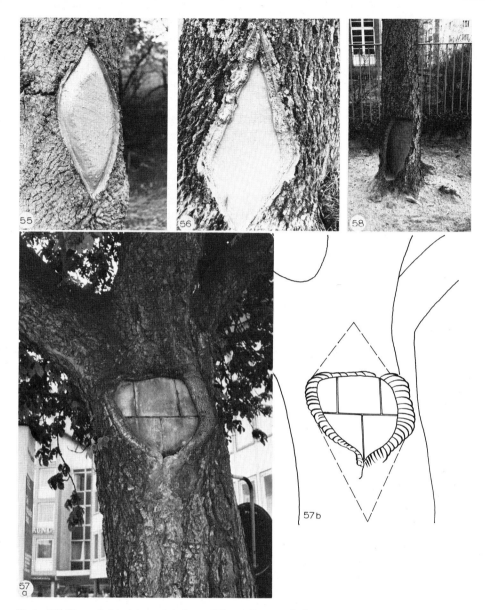

Plate 55. Wound shaping; good shape (Photo Bernatzky).

Plate 56. Bad callusing, because wound shaping below is too short (Photo Pessler).

Plate 57. Bad callusing caused by bad shaping; correct shaping shown by dotted line on drawing (Photo Stummer).

Plate 58. Bad shaping; sap flow from above interrupted (Photo Maurer).

186

wood and bast. When the overgrowth joins over the wounded area the tissues fuse and the cambia once more unite into a meristem layer. When radial growth occurs, that is prior to May since radial growth occurs chiefly from May to July, tree wounds heal intensively. Wounds made after July will heal little during the subsequent months (Neely, 1971).

7.2.1.4. Wound treatment. This is neglected in most cases. The reason for many a trunk rotting and for the dying of whole trees is found in the neglect of the cut wounds. Nature takes care that the organism of the tree is protected all round against bacteria, fungi, etc., by bark and the tissue beneath it just as skin protects human beings. Therefore all types of wounds, no matter how caused, must be expertly treated immediately. It does not matter whether the cause is a car crash, a bolt of lightning or just the saw cutting off a branch. The aim of treatment is rapid creation of the protective callus which reseals the wound. In general, callus grows 1—2 cm per year depending on type, age and nutrition of the tree. Knowing this, it is easy to calculate the time required by the tree to close up the wound completely. This is of interest in insurance cases involving damage claims since, throughout the time of healing, the tree has to be checked continuously and given further treatment.

Furthermore, the rules already stated should be observed, i.e. smoothing of the wound edges with a sharp pruning knife and enlarging the wound to a pointed oval shape. Then the wounded area is completely coated with shellac or disinfected with denatured alcohol. Also, the tools used have to be disinfected by boiling or dipping in 70% alcohol or in a chlorine solution. The wounded area should be closed with a good sealing material (Lacbalsam) against fungus infestation, but a single treatment is never sufficient. Adequate results will be obtained only by regular inspection of the wounded area and by additional treatment once or twice a year. These repeated checkups take care of blisters, cracks, etc., in the wound dressing and other damage. Before repeating the treatment, the loose parts of the dressing should be brushed off. Already developed callus should not be covered with wound dressing material. An overall good supply to the tree of water and food and protection against pests help to wall over the cut quickly.

7.2.1.5. Tree wound dressing. Ever since trees have been prunced — and that has happened for thousands of years — man has thought about protecting the cuts and speeding up their healing. Widely varying compounds have been applied, including loam squash and cow dung, wax and synthetic resin, with varying degrees of success. A good wound-healing compound must fulfill the following functions:
disinfection of the wound area,
toxicity against parasites,
harmlessness for living tissue,

support for the callus growth,
porosity for evaporating moisture,
protection against intruding moisture,
protection against cracking of the wood,
elasticity.

It is very difficult to unite all these properties of the living bark in an artificial product (Shigo and Wilson, 1971, 1972). Pirone (1972) goes into this problem and investigates most of the various wound-healing compounds used so far; he confirms that not one of them meets all the requirements.

Grafting wax, though it is a biological product, melts easily and has no durability. Other compounds were originally composed for use on dead wood. Orange Shellac belongs to this group; it is also used for the first coat of painting. It will protect the wound edges against drying up and disinfect by its alcohol content, but the effect does not last. Oil paints, likewise developed for surfaces of dead wood, contain substances harmful to living tissue. Then there are asphalt mixtures and lanolin mixtures. The latter are not toxic for living tissue. Pirone suggests the following composition: Two parts of lanolin, thinned with one part of raw linseed oil, to which 0.25% of the total mixture of potassium permanganate solution is added. Because the permanganate is not soluble in the lanolin—linseed-oil mixture, it must be dissolved in a small quantity of acetone before being stirred into the mixture. Also used are varnish—mercury mixtures (toxic!) and bordeau-paste which unfortunately delays callus growth and decomposes speedily. Besides these, some prefabricated wound-sealing compound produced by different firms are on the market. Many of them impregnate only the wounded tissue, causing it to dry up and die so that parasites can get in all the more easily. Besides this, they are more or less impenetrable so that no condensation water can get through. They are, after all, coatings for dead wood, not for living tissue. Shigo and Wilson (1972) tried out several wound dressings: asphalt, shellac and polyurethane varnish. They wanted to find out what happens inside the tree. Their findings were not positive. "Bacteria and non-decay fungi invaded all the wounds. Within each tree species there was no statistical difference at the 95% confidence level in the amount of discoloration associated with the wounds. These results indicate that the wound dressing had no effect on invasion of microorganisms or on the processes of discoloration after one year".

Recently a German product was brought on the market called "Scheidlers Lacbalsam". It is a synthetic dispersion emulsion. Judging from its chemical analysis, in structure and function it is an artificial bark with all the qualities of the living bark. Most of its substances are tissue-related. Like natural bark, the Lacbalsam coating forms a weatherproof shield outwards, allowing at the same time for plant respiration and transpiration. No condensed water can gather under this skin. Because Lacbalsam dried elastically it conforms to seasonal and weather-caused deviations in the wood expansion without tear-ing. This elasticity prevents destruction of the protective coating by tension and a renewed exposure of the wound tissue. So far, experience has shown the climatic durability of Lacbalsam to be about 5 years. Within this period, a new infestation of parasites or fungi into the tissue wounds need not be feared as Lacbalsam is not a nutrient medium for them. It can therefore be used for treatment of all tissue and bark injuries, for cancerous deformities and ulcers, as well as for grafting work on trees and shrubs. It does not penetrate into the tissue, but adheres to the surface like a rubber-elastic coating. No cracks will occur in this surface because Lacbalsam does not harden the tissue. Lacbalsam is painted on with a brush or spatula as soon as the disinfecting liquid has dried up. It should not be used on days with frost or heavy rain. For larger injuries "Lacbalsam-Vlies" can be used. Matthews (1974) reported at a conference held at Wye College, that Merrist Wood College together with Surrey University had had satisfactory experience with Lacbalsam. In Western Germany it is used very often with eminent success.

The pH-value of Lacbalsam is 7. In conjunction with distilled water, which in practical terms corresponds with rain water, a slight rise of the pH-value to 8 occurs (according to Lötschert, 1975, Prof. of Biology, University Frankfurt/Main, hitherto unpublished experts' report). This pH-alteration moves within a range which must be called very favourable physiologically. Since in immission areas of cities and overcrowded regions an acidification on the bark of trees is noted, the slight alkaline reaction is very advantageous. It counteracts excessive acidity and can therefore be used with special benefit in immission regions.

Deciduous trees and conifers may likewise be treated with Lacbalsam. The resin of the conifers alone is not sufficient protection against injuries. Painting old neglected wound areas on trees with a wound dressing is quite useless. The cuts must be sealed at the time of pruning and remain so until they are covered with new tissue. Therefore continuous inspection is needed and if necessary a renewed wound dressing. Newly developed callus need not of course be covered.

7.3. Peeling bark

Often whole areas of bark are peeled off or loosened by lightning or by colliding vehicles. If the detached bark is still partly joined to an unaffected zone, and if the former is still moist inside, it may be nailed immediately back onto the wood of the trunk, using rust-proof nails (aluminium). There is then some hope that it will "grow on" again. Then, as already stated, all that remains to be done is to cut the edges of the bark smooth and to coat the seam with Lacbalsam.

If the bark and the exposed wood are already dry, all of the loosened bark must be cut back to where it is still attached. Small islands of bark do not contribute to faster recovery; they can be removed. It is extremely important to protect wood exposed in this way with Lacbalsam. This coating must be inspected for cracks several times a year.

Bruising of the bark at the point of transition between trunk and roots, the root collar, is particularly dangerous. If these injuries are neglected, the vertical stability of the tree may be seriously impaired due to decay at the root collar.

7.4. Bridging of trunk wounds

Large wounds can be cured more quickly by bridging. This requires one-year-old shoots of the same species, slightly longer than the vertical diameter of the wound. Using a sharp knife, the shoots are given a simple grafting cut on both ends (as for grafting behind the bark) with the cut surfaces pointing inwards (Plate 59). The length of the prepared shoots is now the diameter of the wound plus the length of the two cuts. Above and below the trunk wound, a vertical incision is made in the cambium the length of the grafting cut; the bark is carefully detached and the shoot pushed in with the cut surfaces facing the wood. To obtain a better growing together, the uppermost hard bark tissues should be removed from the shoots exactly at the place where they come to lie beneath the cambium. The shoots may be nailed down with rustproof nails, but it is better to tie them with raffia. Now all cuts must be coated and made waterproof. The use of cold-fluid grafting wax or a special plastic material (Lacbalsam) is most appropriate. The more shoots that are used for bridging, and the closer they are placed, the faster will be the healing process of the wound. The best time for this work is early Spring when mature annual shoots that are not yet sprouting are available and the bark can easily be detached.

The following is another method to save and strengthen a young but precious tree suffering from root damage or stunted growth. One or more pencil-strong plants of the same species are planted as close to the tree as possible and left growing for a year. In the following year in Spring, as soon as the bark becomes easily detachable, the trunk is given a T-cut as for inoculation but in the reverse direction so that the cross cut is done below the vertical cut. Now the bark is carefully lifted from the trunk and the shoot, trimmed at an angle, is pushed under, bound up and sealed. In most cases the cross cut will have to be slanted so that the surfaces of cut and shoot join tightly. Basic requirements for these little surgical operations are a sharp knife and quick, careful work to prevent the growth-promoting cambium surfaces from drying up. It is well known that these methods of bridging wounds and planting young nursing shoots are by no means new; yet they are applied very rarely. Perhaps they may in some cases help to save a valuable tree and spare it from the axe.

190

Plate 59. Bridging of trunk wounds, "tree-nurses" (Photo Bernatzky).

7.5. Bracing

Some situations demand the mechanical support of trees and also their bracing. It should be done in many cases including V-shaped crotches, split crotches, twin trunks, frost cracks, cavities and extending branches (Plates 60—62).

Early efforts at tree bracing involved obsolete methods including the use of long rigid iron bands and bars put up high into the tree crowns, iron collars placed around split crotches, chains to hold limbs in position and various systems of wires and home-made cables. The harmful effects of such treatment can be seen everywhere even today. The iron bands and chains strangle the growing trunk by shutting off conductive vessels. The result is thick bulges, especially over the strangulations. The bark underneath the iron band dies. It will be overgrown gradually but without the bulges ever joining each other. After at most 10—15 years, these branches break off at the weakened points. Even if a tree should rid itself of such strangulations, which happens very rarely, it will never recover. The danger of breaks will always be there. Old rusty iron collars must be removed with the greatest care. They have very high tension and loosening them might cause explosive bursting in the

Plate 60. Split trunk caused by V-shaped crotch (Photo Bernatzky).

Plate 61. The rotten crotch (of the same tree as Plate 60) (Photo Bernatzky).

manner of projectiles. After the removal of the iron band, some help can be given to the tree by making incisions across the bulges in a longitudinal direction right down to the cambium to further new callus growth and later the development of new conducting vessels.

Today screw rods are used and fastened right through the trunk for the support and stabilization of trees (Fig. 68). The rods have a diameter of 10—20 mm and are continuously threaded without interruption. In order to attach them, a hole is drilled right through the wood (with an auger or a drilling machine), the hole being 1 mm bigger than the diameter of the screw rod (Plate 63). The rod is put through the hole and fastened at both ends with nuts and washers which, after the removal of bark and bast, can be

Plate 62. Froist crack, untreated (Photo Bernatzky).

screwed right onto the pretreated plain wood. Now all cuts are disinfected and treated with Lacbalsam in the same way as explained for other trunk wounds. As many screw rods are installed, 30—80 cm apart, as are necessary for stability. If possible they should not come exactly beneath each other in order to prevent the same vessels from being hit all the time, and to further trunk stability (Plates 64—66). Whenever two adjoining parts of a trunk are to be linked in this way, the threaded rod has to be covered with a rustproof protective pipe at the gap between the parts (Fig. 69). Bolts might be used instead of screw rods but screw threads would have to be cut in at their ends, which means more and special work. Prefabricated threaded rods save time.

In the pits of pointed crotches in trees, the bark and cambium get pressed and destroyed in the course of growth; thereafter, splits occur. Water enters here, which freezes in Winter and splits the trunks. Fungi also get into the trunk along with the water. This damage is externally invisible and some day the tree suddenly breaks apart. Therefore these splits call for the special care of the tree surgeon. Treatment consists of thorough, lasting draining of the pits, with pipes if needed, and of bracing with screw rods and elastic cable anchors (page 196). Already decaying wood must be removed carefully and

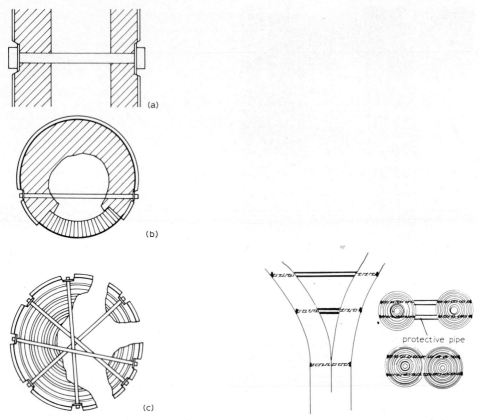

Fig. 68. (a) Bracing with screw rods; (b) "false filling"; (c) keeping trunk remnants together (screw rods at different levels).

Fig. 69. Securing a trunk with screw rods.

the wounds treated afterwards (page 186). On split trunks, all loosened bark should be removed and the edges cut smooth with a sharp knife. In order to join the parted surfaces, it might be necessary to use a block and tackle before the screw rods are installed, but that involves the danger of further splitting. Such V-shaped crotches should be treated as early as possible in the juvenile state of the tree. They develop wherever two or more leading shoots are present and only one should be allowed to keep on growing (Wagner, 1968, 1969). Some examples of obsolete bracing are shown in Plates 67 and 68.

7.6. Frost cracks

These are caused by differences of tension in the trunk, developing with the increasing gradient of temperatures between north and south sides. This

Plate 63. Hole-drilling (with a hand auger) for rigid bracing (Photo Pessler).

happens particularly in the Spring when the sun warms the trunks during the day. Then the trunks burst with a big bang, sometimes along the whole length. One should wait until the cracks close again by themselves, then screw them together with rods or bolts as mentioned. This is necessary because the split parts will not grow together, although they get closer, and will burst again in the next Winter. This is precisely what must be stopped by the screwing together. A prior condition is that bark and wound treatments have already been given and drainage, by a steel pipe with a diameter of 15— 20 mm, has been installed. Remaining pieces of loose bark, small stones, etc. should be carefully removed before screwing together, otherwise it is not possible to close the crack completely. Loosened bark must be cut back to the firm, healthy bark. If nothing is done at all, decay will set in along the whole length of the crack. Possible precautions against frost cracks are

196

Plate 64. Arrangement of the screw rods (Photo Pessler).

boards put on the trunk's sunny side in Winter, or painting with whitewash, or wrapping the trunk with burlap.

7.7. Cabling

As an alternative to rigid bracing, branches or whole trees in danger of breaking can be more flexibly supported by steel cables. Properly installed, they become quite invisible in the tree crowns. They will neither press nor crush any bark as the iron band does, nor any vessels. Linking several parts of the trunk or branches in this way secures the crown not rigidly but flexibly. The branches are able to sway in the wind as before and at the same time escape some of its force. If a branch should break off despite the cabling, it is held by the cable and cannot fall into the street or park (Plates 69 and 70).

Heavy-duty steel cables 4—15 mm in diameter (usually 8—12 mm) are

Plate 65 and 66. The nut of the screw rod (Photos Pessler).

needed for cabling; also screw rods with washers, hexagonal nuts and ring
nuts, thimbles and rope clamps. In Western Germany this anchoring method
has become widely accepted. The anchor is installed as follows. The screw
rod is put into the predrilled hole and fastened on the outside with washer

Plates 67 and 68. Obsolete bracing (Photos Bernatzky).

Plate 69. Good and correct cabling with screw rod anchors (Photo Pessler).

and nut and on the opposite end with washer and ring nut. Through the eye of the ring nut a thimble is placed to take the cable and this is fastened tightly with 3, 5 or 7 cable clamps. The same is done on the other limb. If two branches are to be linked in this way the anchors, if possible, should continue the direction of the cable in the wood of the branches. To achieve this one or both branches may have to be drilled at a sloping angle, which is not always easy. All wounds and cuts should, of course, be protected with Lacbalsam. It is a point of controversy whether more than one cable may be fastened on one anchor (Wilks, 1969; Beeching, 1969, 1970). If the anchor is only a lag hook drilled in, and not going through, the trunk, two or more cables would tear it in different directions. The hook would probably loosen and widen the bore wound, if it did not break off before that. However, an anchor put straight through the trunk and fastened on both ends with washers and nuts or ring nuts, directly conducts the tension of pull on the ring nut. The screw rod will not be affected by the different tensions. In this case several cables can safely be attached to one anchor (Figs. 70—72). In Western Germany innumerable anchors have been installed in this way without a single complaint. The use of lag hooks, hookbolts, etc. for attaching the cables (as used in the U.S.A. and in other countries) cannot be recommended. Their bearing capacity is limited and cannot be compared with that of screw rods with ring nuts put right through the wood as mentioned.

Usually the anchors are placed at about two-thirds of the distance from the tree trunk to the end of the branch, so that the uppermost one-third of

Plate 70. Good and correct cabling with screw rod anchors (Photo Stummer).

the branch can sway freely in the wind. To determine this point of anchorage requires the special skill of the tree surgeon and can be the most difficult task of cabling. It should be placed at the statically correct point which differs from tree to tree. The tree surgeon must also judge the strength of the flexible cable and make a calculation about the weight of the wood, rain, snow and ice and about the wind pressure. In addition to accurate calculation a great deal of practical experience is needed. It is also debatable whether the cable should be held absolutely tight or slack to a certain degree. Extremes to one side or the other should be avoided by all means. Continuous tension must be prevented. The cable should come into action only with the force of wind, rain, snow and so forth, but must not slacken too much either. This again is a matter of the tree surgeon's experience and practice. Each tree is a different living individual and its troubles have to be answered in terms of the peculiarities of the individual tree, its species, age,

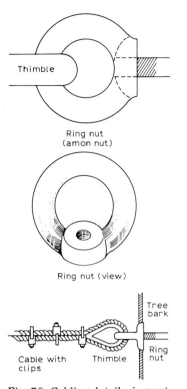

Thimble

Ring nut
(amon nut)

Ring nut (view)

Tree
bark

Cable with   Thimble   Ring
clips   nut

Fig. 70. Cabling-detail; ring nut.

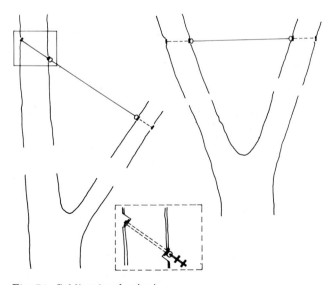

Fig. 71. Cabling (anchoring).

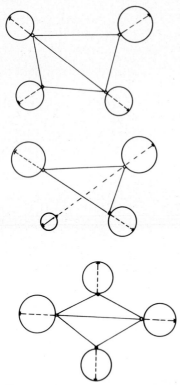

Fig. 72. Cabling of trees by screws rods and ring nuts.

quality of wood, structure, health and the soil it grows on. It has proved to be good practice to slacken the cable as much as it sags in unloaded condition: 2—3 mm for each meter of length. Of course the quality and condition of the cable have to be considered for safe loading. Beeching (1970) gives the following recommendation. "It is largely a matter of experience, and one step towards gaining this experience is to take a good look at the trees on a windy day. The rise and fall of a lateral branch is easily determined by throwing a weighted line over any given point and lowering the weight to the ground. Wait until a strong gust of wind comes along and measure the distance which the weight is lifted".

Cables and anchors must be periodically checked, preferably every year. After about 15 years, however, they have to be inspected very thoroughly, not only from the ground but directly in the tree. Some trees may be saved by bracing and cabling against storm damage, especially along roads and streets with heavy traffic. The danger of injury to people and damage to property might thus be eliminated (cf. Wagner, 1968; U.S. Dept. Int., 1963).

7.8. *Cavity treatment*

Cavities are always the result of neglect or faulty care; either the injuries are caused by natural phenomena (lightning, storm, frost-crack) or by man (collison of cars, incorrect pruning, wrong wound treatment). Obsolete cavity treatments are shown in Plates 71—74. Actually the aim of tree surgery must be to prevent the development of such cavities by immediate and correct wound treatment (page 186) and to stop decay at the start before it causes cavities.

Plates 71—74. Incorrect, bad cavity treatment (Photos Bernatzky).

Cleaning the cavities. First, all decayed wood should be scraped out of the cavities. Suitable tools for that purpose are the chisel and the gouge with a curved edge, both in various sizes; a mallet of wood or hard rubber; a sharp knife and an auger (page 216). There is also machinery such as the automatic gouge, automatic drill etc. It will never be possible to remove all the infected wood completely because the stability of a tree would then diminish so much that it might be lost completely. But with due regard for this fact, it will always be better to remove as much of the decayed wood as possible. However, Pirone (1973) quotes Wilson and Shigo according to whom trees are capable of walling off decay that follows wounding. After a tree is wounded, a wall or barrier begins to form. The wall separates tissues present at the time the tree was wounded, and which may be invaded by decay-producing fungi, from tissues that form after wounding and remain free of decay. Wilson advises leaving some of the decayed wood prior to filling a tree cavity, thus allowing the protective wall to remain.

Sometimes the opening of the cavity has to be widened for the removal of decayed wood. If callus bulges have already developed at the edges, they should be spared as much as possible as the conducting vessels in those bulges also serve the subsistence of the tree. Should a deep incision into the callus bulge be unavoidable the crown must be reduced so that the balance is maintained. Water must not be able to stay in the cavity. If accumulating water cannot flow off naturally at the lowest point, it has to be drained there by drilling a hole into which a drainpipe is installed (Fig. 73).

Should the cavity reach into the root areas (Plate 75), decayed wood has to be removed from there too. In this case, care must be taken that no stagnating moisture develops but that all water can flow off instantly. The ground may also need to be dug up and loosened to ensure lasting permeabil-

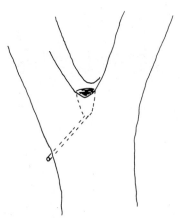

Fig. 73. Drained water pocket in base of forking branches.

Plate 75. A properly treated tree; root stock cavity! (Photo Maurer).

ity. After disinfecting and coating all surfaces with Lacbalsam, the part of the cavity beneath the ground surface is filled up with coarse gravel (diameter 30 mm) up to 5 cm above ground level. In the cavities of linden trees internal roots are found (Plates 76—78) growing from the onset of the crown downwards. They grow into the accumulating litter and humus at the bottom of the hollow trunk, thus supplying the crown with nutritive substances and strengthening the stability of the tree. Therefore they should never be removed or injured in any way.

Forming the opening of cavities. Plates 79—81. These openings should also have a pointed oval form (Fig. 74) as already mentioned in the section on pruning and wound treatment (page 180). The flow of the assimilates out of

Plates 76—78. Inside roots in hollow linden trees (do not remove!) (Photos Bernatzky).

the foliage crown can thus easily reach every single part of the cavity edges. The edges must be cut smooth with a sharp knife, particularly after work with a power saw which tears the cells to pieces, and painted with shellac or Lacbalsam.

Disinfection. All parts of the wood laid open, but not the living tissue (!), must be disinfected with Xylamon-brown, carbolic acid, copper sulphate (0.5 kg/15 litres water) or mercuric chloride (21 g/21 l water). The latter two disinfectants are very poisonous. As soon as the disinfectants have dried up, a wound dressing is applied to the whole cut surface. Lacbalsam produces the best results. After the work is finished all implements must also be carefully disinfected to avoid further infections such as DED.

Plates 79—81. Correct wound shaping and cavity treatment (Photos Pessler).

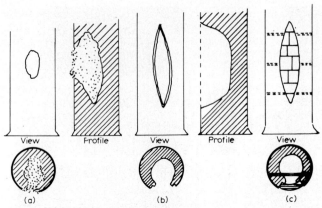

Fig. 74. Cavity treatment. (a) Rotting cavity; (b) opening, cleaning, disinfecting and sealing of the cavity; (c) bracing and applying a "false filling".

Reinforcement of cavity-weakened trees. Often cavities are so big that they seriously impair the stability of the tree. The installation of screw rods can give some help (page 192). They are attached across the cavity of the hollow trunk according to the stability of the tree, but it is necessary to make sure that the surface between the cavity edge and the screw rod (the wood bridge) does not break off in a storm or similar occurrence. The breadth of the wood bridge depends on the remaining thickness of the wood, on its specific qualities and on the whole situation. It is useless, therefore, to specify an exact distance from the cavities' edge. The rods should never be placed exactly above each other but always somewhat staggered in order not to disturb the same conducting vessels each time.

A vertical reinforcement in a cavity (as sometimes suggested) by vertical bars or something similar is useless. In order to carry the weight of the tree crown their dimensions would have to be far larger than the actual opening of the cavity. Also, a filling with concrete or other material in no way improves the tree's stability but may, indeed, reduce it still further (lopsided trunks). The weight of the crown can be more effectively supported by other means (page 196).

Filling up a cavity or not? Opinions differ on this question. What is the sense of a filling? It cannot remove decay or embank it. Instead it offers optimum living conditions to pests and vermin in the interspace between wood and filling, which can never be tightly closed. The constant temperature there, as well as the lack of ventilation and sunlight, favour their increase and they can do their work of destruction without being noticed. Rigid fillings like cement do not join in the movements of the tree in the wind, nor do they take part in the physiological processes, e.g. contraction of the wood at noon due to water stress and expansion at night. In fact the expansion-coefficient of the filling material hardly corresponds with that of the tree wood. This results in counteracting forces, not suitable for either the trees or the filling material. One possible reason for filling would be to give the growing callus a support which it can overgrow so that some day the new bark growth will seal the opening completely. Without such help the new bark grows with an inward bulge and the two edges may never be united, particularly at large cavities.

Open cavities. Plates 82—84. Thus the closing of a cavity has no effect on the tree's stability whatsoever. If it has been secured with screw rods as described, an open but correctly treated cavity will not endanger the tree's stability in any way. The rolling in of the new callus towards the inside of the cavity reinforces the stability of the trunk better than new bark over-growing a filling. The decisive argument in favour of open cavities is given as follows: inspection of the tree is far easier in terms not only of the work of the tree surgeon and his assistants but also of the development of the tree itself, stop of further decay, the quality of the wound dressing and so on. Very large cavities may be secured with wire netting so that children cannot light fires in the hollow trunk or use the screw rods for climbing.

Plate 82. Open cavity (Photo Bernatzky).

Plate 83. Open cavity (Photo Bernatzky).

Filling of cavities. Other authors and tree surgeons recommend the filling of cavities. In former times cement mortar was used as the cheapest material. Asphalt mixtures, cork products, rubber blocks (Pirone, 1972), magnesite compounds (Fenska, 1964) and other materials have also been used, most of them protected by patents, actual or pending, but these products hardly meet the requirements for a good filling material which should be
durable and unchanging in heat and frost,
flexible to respond to the movements of the tree,
elastic to fit tightly into the cavity,
waterproof, so as not to stimulate decay of adjoining wood,
and harmless to the living tree.
For the technical details of such fillings see Pirone (1972), but once more the difficulty, and sometimes the impossibility, of inspecting filled cavities must be pointed out. Cement fillings are especially unfavourable because of their rigidity and lack of durability. They favour growing conditions for invading fungi by shutting out air currents and sunlight. The man who wants to place such a filling should be particularly experienced in the techniques involved as well as thoroughly familiar with the physiological requirements of a tree and its living conditions. It is better not to make a filling than to carry out a wrong one!

Plate 84. Bracing and cabling of an old linden tree ("Prior linden" near Hagen/Germany) (Photo Maurer).

The use of polyurethane for filling of cavities. Lately the use of poly-urethane has been recommended as filling material (Palmer and May, 1970; King, 1971; King et al., 1970; Banukiewicz, 1973). Its application is inexpensive, simple and saves time. A disadvantage is the fact that no further inspection of the treated cavities is possible. For the application of this material the cavity has to be cleaned and disinfected, the opening must be formed, wound dressing applied and screw rods installed if needed. About 1 cm (1/2 inch) around the edge of the cavity loose bark has to be scraped off and removed, and a weather stripping placed which, when installed, will form a tight seal. Aluminium foil is now sprayed with aerosol silicone, placed over the cavity and tied securely. An opening is left at the top to pour in the Vulta-Foam 15-F 1802 (General Latex and Chemical Corp., Cambridge, Mass., U.S.A.) mixed according to regulations. In a few minutes the substance will start to foam and fill the cavity. The heat briefly generated in this process will not harm the cambium according to King et al. After another ten minutes the mixture has hardened. Should there be too little of the material, more can be added now. The aluminium foil is removed from the hardened foam and the edges trimmed with a simple jack-knife to the shape necessary and desired. The foam surface should not cover the cambium layer on the cavity edges. The whole surface is coated to the edge of the wood with a mixture of auto body filler and hardener and subsequently with tree paint. On average, the whole procedure takes about 20 minutes for a medium sized cavity. The air temperature for this work should be above 65°F = 19°C. According to King et al., Vulta-Foam is resistant to microorganisms.

False fillings. Plates 85—87. If for some reason (aesthetics) the opening of a cavity must be closed, this can be done without a complete filling. For this the sound wood is chiselled away around the edge of the opening about 10—15 cm deep (Fig. 68) and a Rabitz-netting nailed on. A 1 : 3 mixture of cement is now brought in onto the netting, 10—15 cm thick. After the cement is dry, everything gets a coat of Lacbalsam. If the opening is too large to be covered in one piece, the cement can be mounted in layers of 15—20 cm each, held apart by threefold roofing paper. The maximum dimensions of the opening of a cavity must not exceed 30 cm width and 50 cm height. In a similar way a cavity can be closed by a sheet of tin cut to the exact size of the opening. The older the tree, the bigger the cavities — and the more difficult will become the fillings. One should therefore abandon them and, after cleaning and disinfecting, leave the cavity open (Plate 84). The success of a filling not only depends on precise practical work, but requires long and detailed experience, immense knowledge of biological and physiological factors and good judgment. This work, in other words, is definitely only for the most highly skilled expert.

Supporting structures. Plates 88—90. Some trees, often older than 800—1000 years, exist only as relics of the past in the form of solitary bark strips with crowns mounted on top of them, or only the remnants of crowns. Cavity filling would not help at all in these cases, for everything has to be kept open so that the fungi have no favourable climate for their work of destruction. Stabilization of the crown is quite difficult. First, screw rods have to be installed in order to keep together the trunk remnants. Extended branches should not be supported by underlaid boards of wood, but by double props of pipes (Fig. 75). Underlaid props of wood or metal lead to blockage of the sap flow in the living branches just like the obsolete iron

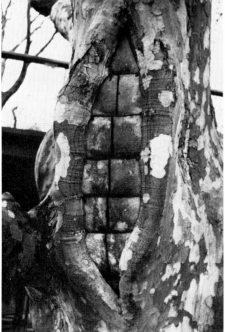

Plates 85 and 86. Good "false fillings" with expansion joints (Photos Maurer).

Plate 87. Unclosed cavity protected with a grating (Photo Maurer).

bands used for bracing. The branch to be supported is bored through horizontally, a screw rod inserted which is fastened at the ends on two metal
tubes laid in concrete in the ground. In order to allow the branch to respond
to the wind, other pipes are inserted into the supporting pipes (telescope-

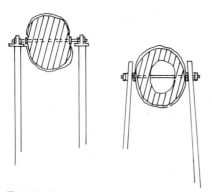

Fig. 75. Supporting props for old trees.

Plate 88. Telescopic props (Photo Bernatzky).

system) and at their upper ends the bolt is attached. Its thickness has to be determined on the basis of the weight to be carried, the thickness of the remaining wood and other local peculiarities. This work demands much practical experience in order to avoid crushing the bark and conducting vessels. As with all tree surgery, pruning may be necessary to improve the stability and aid the physiological processes. Plates 91—95 show several examples of tree surgery.

Tools and equipment for tree surgery

Good tools and adequate equipment are necessary for tree surgery. Generally the tools required come from other industries such as, for instance,

216

Plate 89. Rigid metal prop (Photo Maurer).

forestry or other woodworking professions. Heavy work is nowadays carried out by machinery such as chain saws, from the light and medium weight up to hydraulic and pneumatic saws. Winches and other special machines like stump cutters, brushwood cutters, hoppers and transporters belong to the equipment of large firms in the field of tree surgery.

However, many hand tools still have to be used such as hand saws, cross-cut saws, small pruning saws, pole saws, axes, loppers, two-hand pruners, secateurs, pruning knives, hand augers, chisels, mallets, wire cutters, draw vice tensioners, mattocks, hooks, slashers, etc. (Plates 96—99). Work in a tree crown can be carried out in different ways; from hydraulic platforms, from ladders or just by climbing the tree (Plate 100). For the worker climbing high trees, a special safety outfit is required. Existing equipment includes safety harnesses, life lines, strops, pole-belts, climbing irons, climbing bicycles and ropes. For more details concerning tools and equipment see Bridgemann (1976).

Plate 90. Rigid props at the "Kasberg-linden" tree (1000 years old). The wooden props will be removed after the concrete of the metal supports has hardened (Photo Bernatzky).

7.9. Protecting trees against damage from construction work

It is an old experience that after the completion of construction work, many trees in the neighbourhood die. Often the damage is only noticed years later and is the result of direct or indirect interference with the basic requirements of the trees (Plates 101 and 102). Some suggestions follow for counteracting such damage, especially in the root area. It should not be forgotten, however, that these are mainly makeshifts and not successful in all cases. Much depends on specific local circumstances. At any rate they may be the necessary means of tree survival.

Plate 91. (a) Before treatment; old filling will have to be removed and the cavity cleaned. (b) Later, the cavity will be left open and supported only by rigid bracing (Photos Pessler).

Plate 92. (a) *Taxodium distichum*, before treatment. (b) After treatment; the cavity will not be closed (Photos Bernatzky).

Plate 93. (a) An old beech ("Suentel-beech") in Bochum. (b) Bracing this old beech (Photos Maurer).

Plate 94. (a) Some hundred years old "step-linden" in Schenklengsfeld (Hessian). (b) A part of the tree, before treatment. (c) Later. (d) Another part, before treatment. (e) Later (Photos Pessler).

Plate 95. (a) The 1000-year-old linden tree in Kasberg. (b) The tree in leaf. (c) **Screw** rods in the tree (Photos Pessler). (d) The saved tree (Photo Bernatzky).

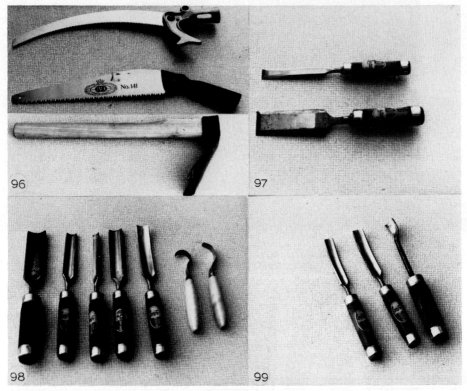

96 97 98 99

Plates 96—99. Tools for tree surgery (Photos Bernatzky).

7.9.1. Filling up the root areas of trees

On principle the roots of trees should never be filled up with earth, not even with humus or topsoil, as the explanations about soil and its micro-organisms and about the roots show. Usually the damage caused by the fillings is only noticed years later when nobody can explain trees being sickly, their growth diminishing, single branches dying, the crown thinning and diseases attacking the tree — all this caused by the overfilling of the root area.

Occasionally filling up may be inevitable and this leaves only two possibilities: either to take away the tree at once or to wait until it dies after a few years. We are able to moderate the severity of the effects by technical means, but it is important to know that despite all precautions filling up will continue to be risky, and the more so when the necessary arrangements are only half carried out or even less. Trees are living individuals and even trees of the same species may react quite differently. The following measures, which are aimed at moderating the damage done by overfilling, can never be more than an outline which must be modified according to the tree species, location and environment.

Plate 100. Ladder for work in the tree crown (Photo Pessler).

7.9.1.1. Technique of applying fills. Firstly the cover vegetation should be removed, turf, plants, leaves and other organic material carefully dug out and the ground slightly loosened. Care should be taken not to disturb the roots of the tree; hence the work should be done manually, not with machinery. Fertilizers may be added before or after breaking up the soil (Figs. 76 and 77). Then perforated plastic drainpipes or porous drain tiles are laid out in such a way that they can serve for watering the tree later. This pipe system, the details of which are shown in Figs. 78—82, should slope about 3% to one side so that surplus water may flow off and not stagnate in the area of the tree crown. The pipe system should cover the whole spread of the crown and preferably go beyond it. The outlets at the lowest point must be connected with the general drainage system, seepage pit or possibly the sewer. If seepage pits are constructed they are to be carefully safeguarded against blocking.

Plate 101. A normal tree becomes broader at the root collar (Photo Bernatzky).

The butt joints of the drainpipes should be protected against blockage by roofing paper, or something similar, and covered with tile chippings. In addition, the pipes have to be covered throughout with layers of stones (Fig. 78). Pipes of clay or some other material (with a diameter of 15 cm) are put

Plate 102. An overfilled tree (Photo Bernatzky).

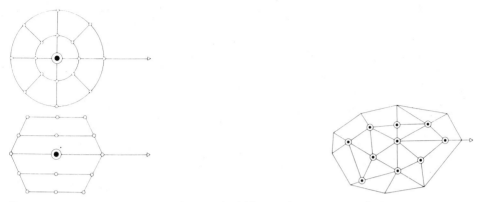

Fig. 76. Aerating system when a fill is to be laid over the root zone of a tree.

Fig. 77. Aerating system for fills applied to groups of trees.

vertically over the joints of the drains and held in position by heaps of stones (Fig. 78). They should rise above the final surface of the fill by about 2 cm to keep surface water from flowing in. During construction work they should be covered with roof paper so that no dirt will get in. On completion of the fill they may stay open or be covered with strong wire netting. Filling the vertical pipes with gravel impedes access of air. The pipe systems of a group of adjoining trees might be connected with each other (Fig. 77).

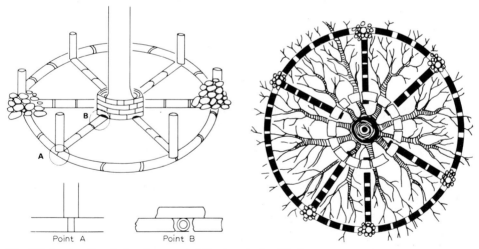

Fig. 78. Laying the drain-pipes round the tree to be filled over.

Fig. 79. Top view of the aerating system. (Figs. 76—79 after A.R.S., 1964.)

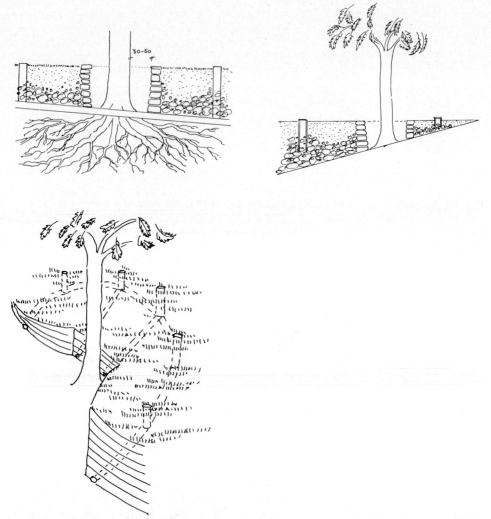

Figs. 80—82. Fills on a slope. (From A.R.S., 1964.)

Around the trunk of the tree a dry wall is erected about 25—50 cm away from it but allowing for the amount of growth still expected. The distance must be larger for young trees, smaller for older ones. Natural stones, hollow blocks or bricks, solid or perforated, serve as material for this wall. Hollow stone blocks should be laid with the openings showing a vertical direction so that water and air can get through unhindered. The stones might be set in lime mortar but open joints must be left between the bricks for the air to pass through. Only the uppermost ring of stones is laid tight in lime mortar. The top edge of the wall should be slightly above the future ground level to keep out water, rubbish and dust from the space around the trunk. Best of all the wall is covered up with wire netting but not filled with stones or gravel for air access would then be impeded. The single drains of the pipe system should reach all the way up into the air space around the trunk.

The whole surface of the original ground level is now lightly covered with a layer of stones (10—20 cm diameter; Figs. 80 and 81). When the height of the fill is 75—100 cm, the layer of stones should be 25—30 cm; with higher fills, which are increasingly risky, the layer of stones can be up to a maximum of 70 cm. Towards the trunk the layer of stones should ascend until it is 30 cm below the surface of the fill. The stones should be laid so that enough air space is left between them. In order to stop some earth falling between the stones later, a piece of glass fibre or similar material is spread over them. Only then may soil be added and planted.

On sloping sites the pipe system need not be installed. Here it might be enough to cover the slope with stones as described and ventilate it upwards with air pipes (Figs. 81 and 82).

7.9.1.2. Treatment of trees which have already been filled over. Trees suffering from fills are generally recognized by the fact that the trunk does not widen at the base. A normal tree becomes noticeably broader at the root collar. Whether trees already filled over can still be saved depends on a number of factors: tolerance of the type of tree to fills in general, height, type and duration of fill, extent of damage already detectable, age of the tree.

As an immediate precaution, air might be blown in by a compressor at intervals of 1—1.5 m down to the former ground level. With porous soil and shallow fill this may be sufficient, especially if it is repeated at intervals. The best remedy, however, is taking away the whole extra layer of soil and installing the above-mentioned aerating system. If this seems too expensive then the drainage system should at least be dug into the ground in narrow trenches and laid there. Alternatively, as happened in Germany recently, fleece wicks (plastic) 50—80 cm long can be inserted in the ground at intervals of half a meter (Plates 103—106). The success of all these methods depends on the amount of air that has been brought to the old roots (Fig. 83), but unfilled pipes are always best.

All the ventilation systems mentioned can also be used for temporary irrigation during drought.

In conclusion it should be emphasized that no tree whatever will stand filling over without disturbance. The younger the tree the better it will adapt to the changed conditions that a filling-over will cause, but the older the tree, the poorer will be the possibility of adaptation! Certainly in most cases poplars and willows, and also perhaps some other trees, will develop new root levels after overfilling, and even hard-wood types such as oaks may do the same, but simultaneously with the development of new roots the old ones die much faster than the new ones grow so that the stability of the tree,

Plate 103. Fleece-wicks for aerating and watering trees (Photo Pessler).

especially the tree with a large crown, diminishes. What that means in densely populated towns and on roads and streets with heavy traffic is well known, and yet the old sin of the gardeners to fill over trees will most probably never be stopped. It is just too bad for the trees!

Particularly vulnerable, and usually killed by overfillings and compactions of the soil, are the following trees, almost all with a lateral root system: birches, beeches, maples, hornbeam, planes, walnut, conifers, some oaks, the Tree of Heaven (*Ailanthus*) and many members of the *Rosaceae* family.

7.9.1.3. Preservation of trees standing in concrete and similar surfaces. Concrete, asphalt and similar surfaces have the same effect on trees as overfillings of soil. The best help for the trees concerned would be to remove the impermeable layers, to air and water the ground beneath the tree crown and

Plate 104. Fleece-wicks inserted in the soil or pavement (Photo Pessler).

Plate 105. A cap for the fleece-wick (Photo Pessler).

Plate 106. Fleece-wicks installed in pavement (Photo Pessler).

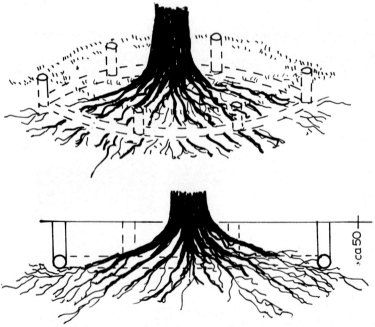

Fig. 83. Installation for aerating and watering of old trees (by plastic pipes or fleece-wicks).

Plates 107 and 108. Soil cover pervious to air and water (Photos Bernatzky).

232

to plant it with *Pachysandra*. This will, of course, not always be possible especially in streets and on city squares. Another possibility is the removal of the concrete layer and replacement by paving stones, compound stones, or hole-and-slit stones, laid out in sand. In this process the ground must not be in the least compressed or consolidated by machinery or otherwise. Alternating iron gratings with big and small paving stones and filling the interspaces with coarse gravel can achieve aesthetically attractive effects (Plates 107 and 108). Should this also be impossible, strips of 1 m breadth (as many as possible) can be cut into the concrete. These should be laid out radially from the trunk towards the outer perimeter of the crown, and one should go round this perimeter. The concrete should be removed and the area paved with stones set in sand (Figs. 84 and 85). Pipes might be inserted at the same time to air the ground and bring in water and fertilizer. Each of the solutions mentioned is less effective than its predecessor, but it is always better to do something to save a tree than to let it die by doing nothing.

7.9.1.4. Other compactions of the soil. Construction vehicles of all kinds, with pneumatic tyres as well as track vehicles, press the air out of the soil and greatly consolidate it. Road builders and similar constructors use special

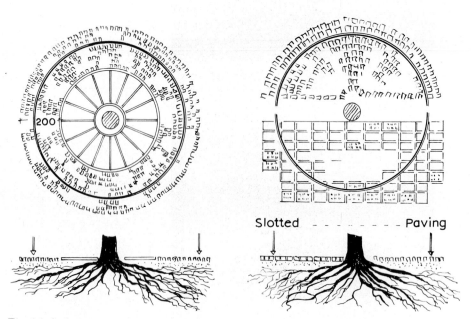

Fig. 84. Soil cover pervious to air and water: iron grating, then cobbled paving.

Fig. 85. Soil cover pervious to air and water: cobbled paving in sand, or special stones (perforated or slotted).

compressing machinery (vibrators) which are very harmful to trees. Continuous walking on the root area by workmen has the same effect, as does the storage of building material, oil and so forth. The best remedy for these damages is to avoid them. On no account should the soil be broken up by machinery since this harms more than it helps. Instead the soil should be lightly loosened with a fork or hoed by hand. The showing of legumes (*Leguminosae*) is conducive to soil porosity. Stagnating water is another form of consolidation. It mostly occurs in positions where clay soils or floating sand is driven over by heavy duty vehicles. It is true that such soils can regenerate themselves but this takes years and meanwhile the trees suffer badly. Should the ground below the surface be permeable, help can be given by drilling deep holes and filling them up with coarse gravel. Care should be taken not to damage any main roots in the drilling process.

7.9.2. Soil removal
Each removal of soil deprives the tree of its basis for nutrition because in the upper layer of natural soil, usually 30 cm, most nutrients and micro-organisms are found. Moreover soil removal damages the roots and weakens the stability of the tree, especially if it is shallow-rooted. Therefore the area beneath the crown should not be touched. This is easy in the case of a one-sided excavation, for instance where there is a deeper lying road, because removal of the soil can begin outside the outer perimeter of the crown. On the edge of a slope, the difference in level may be secured by a retaining wall. As such work proceeds, the soil should be continuously aerated and the tree fed. If a tree is left standing on a mound with the soil all around it removed, there is little hope that it will remain alive, particularly if it belongs to a shallow-rooted species and has a low ground-water level. Installation of irrigation facilities may extend the life of the tree. Removal of soil at close range to trees should never be done with excavators or other machinery because they will tear up or crack off roots as far away as 30—50 cm beyond the edge of the excavation. As damage of this kind is invisible, it will not be treated and the roots will start to rot.

7.9.3. Measures in the root zone of trees
It must be emphasized again and again that the root zone of trees should, as far as possible, be left untouched; however, if this cannot be avoided suitable measures must be carried out.

7.9.3.1. Walls in the root zone. Walls and wall foundations should be interrupted near the tree so that root development is not obstructed. If the whole wall cannot be interrupted, at least the foundations should be. The wall is then set on a beam (Fig. 86) or the whole wall is put on pillar foundations with the beam lying on then. Small structures also should not be built on a continuous foundation in the vicinity of trees but on the same kind of piles.

234

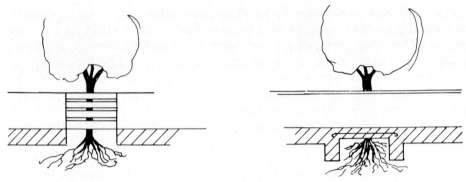

Fig. 86. Protection of roots when a wall is built near a tree.

7.9.3.2. Service trenches, cables or sewers in the root zone. These are among the worst destroyers of trees. As a matter of principle they should always be laid outside the branch spread. It is pointless to state a minimum distance away since trees always lose roots in these cases and there is also

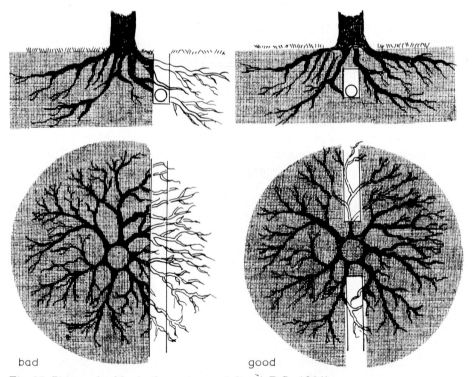

bad   good

Fig. 87. Pipes and cables in the root zone (after A.R.S., 1964).

wide local variation in root systems. However, if it is really impossible to keep outside the branch spread, it is advisable to take the pipes below the centre of the tree because the roots there are least likely to incur damage (Fig. 87). Tunnels must be made for a length of 2 m in front of and behind the trunk. The continuation of the trench should not be dug by an excavator but by manual work only. In addition, strong roots should be tunnelled under. This is, of course, impossible in the case of trees with tap roots; one can only stay away as far as possible from the tap root. Experts consider the distance should be 2.50 m at least. Water drains in the vicinity of a tree should be carefully caulked at the joints because, as is known, roots follow water.

Plate 109. Root curtain (left, stand of the tree; right, building) built up before the construction work (Photo Bernatzky).

Every loss of roots must be compensated by a corresponding pruning of the crown. Watering, fertilizing and general soil care are also necessary for fast recovery of the tree.

7.9.3.3. Treatment of roots ("root curtain") (Plates 109 and 110). If excavations are to be carried out close to a tree, the latter can still be saved if protective work is done, preferably before the start of the construction work. Between the future excavation site and the tree a trench at least 50 cm wide and 1.50 m deep is dug manually, all the roots being carefully exposed and severed cleanly at the side of the trench nearest the tree; sawn edges must be trimmed. The cut ends should point obliquely downwards, and they are painted with Lacbalsam (Fig. 88). On the other side of the trench, towards the future structure, a "fence" is installed made of wooden posts and wire-netting, and hung with sackcloth on the inside (do not use plastic sheeting). Then the whole trench is filled up again with a mixture of compost, ground peat and tree food. Now the tree can easily form new roots at the cut ends. Later, during the excavation work, this root curtain must of course be borne in mind, so that it is not damaged. Depending on soil conditions, the depth of the root system and the skill of the tree surgeon, such a

Plate 110. Root curtain, built up in the course of construction work (Photo Bernatzky).

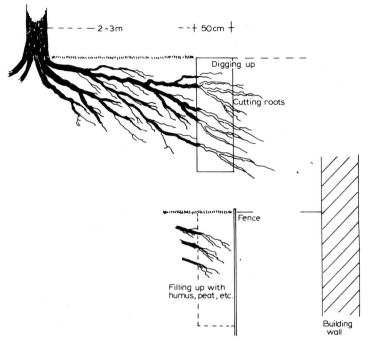

Fig. 88. Root curtain.

root curtain can be installed as close as 2.5 m from the tree.

If the root treatment can be done only during the excavation, the roots must be specially protected against drying out. Also, the excavator should not be used right up to the edge of the root zone, so as to avoid damage to roots far down in the earth. Therefore, root treatment must be carried out stage by stage as the excavation work proceeds.

Whenever possible, root treatment should be done one year before the start of excavation and construction work. Then, no difficulties will arise when using machinery for excavation as long as the root curtain is not destroyed.

7.9.4. Lowering of ground water

This can kill trees. It would be better not to pump away any exposed ground water into the nearest drain, but to bring it back to percolate in a ring of holes, 50—100 cm deep inside the outer branch spread; these should be filled with coarse gravel. This is especially important in Spring, since at this time the water requirement of a single tree amounts to hundreds of litres. If water pressure-pipes are available, a few hoses with nozzles can be hung up in the leaf crown, to spray the foliage in hot sunshine. If a tree perishes during such work, it is a case of wilful destruction.

7.9.5. Isolating of single trees

Sunburn, storm damage, breakage and from cracks endanger isolated trees. Sunburn occurs mainly on the south and west sides of the trunk; beeches and spruces in particular are burned this way. For protection, the trunk is coated all round with loam squash about 3—5 cm thick and wrapped in burlap as used by nurseries. Ropes of straw wound around the trunk are equally effective. Both protective devices should be kept moist all the time so that they can fulfill their function of reducing evaporation. A coating of the whole trunk with Lacbalsam is also very useful. The use of antitranspirants may, but need not, protect isolated tree crowns against excessive evaporation. Storm damage particularly endangers free-standing single trees on wet ground as well as shallow-rooted types. Important solitary trees can be secured by steel cables, reaching from anchors in the crown to foundations in the ground, but this is quite expensive. Cabling to an adjoining tree is always problematic. Extending branches of such a tree are secured against storm damage by steel cable anchorage (page 196). Trees are protected against frost cracks by means of lime coating.

7.9.6. Overhead wires

Trees and overhead wires clash all the time. Either the lines are disturbed by the trees or the trees are mutilated by continuous and expensive pruning. Of course it is even possible to lay out power lines high above forests nowadays, but in all cases the use of underground cables is better.

7.9.7. Measures to protect trees on building sites

Because trees are so easily damaged on building sites and the results of the damage become visible only later, it is necessary to take steps to protect them. It is better and cheaper to prevent damage than to repair it. It is relatively easy to give the right protection (fence, root curtain, etc.) to an existing stand of trees at a building site if it is planned and installed in time. Root treatment should begin 1—2 years before the start of construction. Since in most cases there are no binding regulations for tree protection, protective measures should be included in private building contracts. They should also be included by architects in their specifications for the construction work (Bernatzky, 1974). This involves the following (an example for a contract text):

(1) Trees and shrubs in the vicinity of the building site should not be removed or damaged without the approval of the architect. This includes damage to trunks, branches, tree crown and roots, driving in nails, building brackets, etc. Any significant shortening of branches and roots must be discussed with the contractor and his architect, and should be carried out professionally.

(2) The attachment of tools, switch boxes, pipes and so on to the trees should not be permitted because it causes severe damage.

(3) The root area of the tree (area of ground below tree crown plus 2 m radius measured outwards) must be kept absolutely free from: any passage of vehicles, earth moving equipment etc., machine assembly, site huts, canteens, toilets, piling-up of heavy building materials and fuels of any kind. Exceptions require, in every case, the prior consent of the architect, and an exact specification of the measures required to avert damage.

(4) To ensure effective protection of the root area, either the whole area of the root zone should be surrounded by an immovable protective structure consisting of boards about 15 cm wide and 1 inch thick, and 8—10 cm posts; or a fixed, free-standing (not nailed to the trunk) protective box of planks and posts, whose clear width is four times the trunk diameter, should be erected as before, round the trunk of the tree (Fig. 89). The place where the roots join the trunk should have special protection in the form of sloping boards. The whole remaining root area should be filled with 3/15 cm gravel to a depth of 20 cm and covered with steel plates.

(5) All protective devices (fencing, etc.) should be removed without a trace being left after the works are completed, including any gravel applied.

(6) The trunks of all trees previously standing in the shade, and which were exposed in the course of construction work, should be shaded immediately on the side facing the sun.

(7) Every tree should be covered by penalty clauses, the amount of the pen-

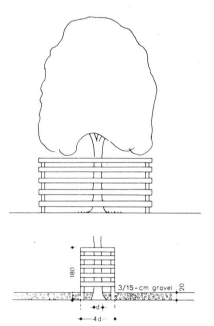

Fig. 89. Installations to protect a tree at a building site.

alty being govered by the thickness and value of the tree. On average, 500—600 dollars are placed on every 10 cm of trunk diameter.

(8) If the tree is damaged, the expenses of tree surgery should be taken out of the penalty. Any additional expenses should also be refunded. If the whole tree is lost, its value is calculated. If these provisions are included in the work specifications, the basis for much irritation and dispute is removed.

(9) On conclusion of the construction work, the soil surface must be cleaned manually by a specialist. Work with a milling tool is not permissible. An application of tree food is now in order. What has already been said about the care of the soil under trees also applies here.

7.9.8. Trees on parking areas (Plate 111)

Parking areas can be divided up by rows of trees. They give shade to the vehicles and screen the place as a whole. They give shelter to adjoining building areas against dust and noise and they may also be important to the town's climate provided they are laid out in the right way. Unfortunately, and just as with all street trees, the environmental conditions in these parking areas never agree with the requirements of the trees. Mainly lacking is a sufficient supply of oxygen and a satisfactory water-holding capacity of the soil. Cars compress the soil. Dropping oil and grease cause further damage, not only because they are harmful in themselves but also by furthering com-

Plate 111. Well planted parking area (Photo Bernatzky).

paction of the ground and by forming a film on the surface preventing aeration and watering. Parking areas with light sandy soil that drain well have much better conditions for trees since there compaction will never be as severe. Plans for new plantings should give full consideration to the living conditions of trees rather than simply letting them adjust in the best way they can. Trees should always stand on elevated borders protected by high curbstones.

Of great ecological importance is the surface-shaping under trees on the planting strips. The sowing of a lawn is unfavourable because it uses up water and nutrients so that hardly anything reaches the roots of the trees (page 28; also Larcher, 1973a). In contrast, planting with small shrubby cover like *Pachysandra* has a very beneficial effect on soil humus and microorganisms; particularly shrubs with no felted root system. Brahe (1971, 1975) in Western Germany questioned all garden and park authorities about covering the soil under trees and their results are given in Table 61.

Any kind of planting though is in danger of being trampled down by people leaving cars, and gets compressed further by being walked over too much. Therefore a cover of small cobbles, hole and slit concrete slabs or flagstones is appropriate and ecologically justifiable as all these covers let air and water through. High kerb-stones, guiding planks and wire netting protect trees and the plantings underneath them. Single wires easily injure people. Well-tried are buffer-stones and steelpipe bows onto which the parking meters may also be attached. For planting and care of trees on parking areas the same rules apply as given for street trees (page 244).

If the paving of the whole parking area is carried out with due regard for tree requirements, they will certainly grow better. Bituminous paving is unfavourable since it burns the foliage by accumulating and reflecting heat into the young crowns, which are not yet large enough to shade the entire area. They also seal the soil, thus blocking out air and water. Stones and slabs laid in concrete are just as bad as fully closed concrete or asphalt layers. The lighter the surface, the more intense will be the heat reflection onto the

TABLE 61
Kind of soil-cover on parking areas in Western Germany (after Brahe, 1971)

Kind of soil-cover	In % of returned questionnaires
Soil covering shrubs	68 %
High shrubs	62 %
Lawn	26 %
Lawn and pavement	20 %
Hole-and-slit concrete slabs	30 %
Gravel	6 %
No covering	64 %

TABLE 62
Kind of soil fixing on parking areas in Western Germany (after Brahe, 1971)

Kind of pavement	In % of returned questionnaires
Bituminous pavement	76 %
Concrete compound stones	64 %
Big and small paving stones	50 %
Cinders	50 %
Flagstones	28 %
Bricks	18 %

foliage. The pavings most favourable to trees are flagstones, slabs, bricks or other stones laid out in sand without sealing the surface. The openings at the joints allow not only air and water to get through but, unfortunately, oil and melted water containing de-icing salt as well. The use of thawing salts on parking areas could be prohibited; under no circumstances, however, should the slope of the surface go in the direction of the tree roots because this will cause salt damage. Another inquiry by Brahe (1971) gives a survey of the paving used on parking areas in West German towns (Table 62). Cincer surfaces are harmful to trees if they contain sulphur traces.

For all plantings indigenous species should be given preference, especially those suitable for dry places and hardy ones not needing much care. In Central Europe these are oaks, plane trees and hornbeam. Trees having fruits like horse-chestnut and tree hazelnut should be avoided because children will always try to knock down the fruits with sticks and stones and so easily damage parked cars. Species developing honey-dew like maple and linden should also be avoided. The distances between the trees should be selected according to the growing power of the different species. Broad crowned trees should be used as much as possible and planted at intervals of 15 m. If the shading effect is to be attained sooner, the distance can be cut in half and every second tree removed at a later date. Old trees on parking areas can be protected only at great expanse both in work and money, for provisions (page 227) have to be carried out such as installations to aerate and water the soil through pipes and sectors of permeable soil covering. Conifers must have at least half of their root area undisturbed. Often underground garages are built beneath parking areas and little soil is left on them. Even here trees need not be abandoned as long as large enough tree containers or tubs are allowed for in the construction. For ecological reasons conifers are generally not very suitable for planting on parking areas, but conditions might be suitable for them on the edges around the place.

8. TREE PLANTING AND INSPECTION

Good and careful planting is the first condition on which the tree's health depends. The planting is for trees what the foundation is for a house. Without a proper foundation the house will not stand for a long time. The same rule applies to trees, especially in the streets of towns. General rules for planting should be known.

Not every tree species can be easily transplanted. Most difficult to transplant are beech, birch, hawthorn, hickory, magnolia, *Nyssa silvatica* (black tupelo), *Oxidendron* (sorrel-tree or sour-wood) and many oak species (Pirone, 1972).

8.1. Technique of tree planting

Planting time
The planting should probably be done in the dormant period of the tree. This period varies from species to species and depends on meteorological conditions such as temperature, precipitation and intensity of the sun's radiation. In temperate zones the best planting time for deciduous trees is between Autumn and Spring, as long as the soil is not frozen; for conifers and other evergreens the time is early Autumn (August, September) or late Spring (April, May). Pirone recommends the following trees for spring planting: dogwood, magnolia, yellow-wood, tulip-tree, American hornbeam, hickory, pecan, white fringe tree, butternut, golden rain tree, sweet gum, tupelo, sassafras and walnut. For birches the best planting time is shortly before the buds burst.

The planting hole
The size of the planting hole depends on the dimensions of the rootstock and on local and soil conditions. In a well cared-for park with normal soil conditions it should be sufficient to make the hole two spade breadths broader and deeper than the roots' spread, and to turn the bottom of the hole one spade deep. The more unfavourable the conditions of soil and site, the more preparations are necessary. Heavily compacted soils may require powerful breaking up or even blasting.

Planting
All injured roots must be cut back with a sharp knife in such a way that

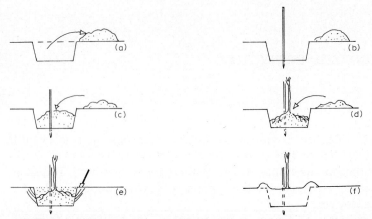

Fig. 90. Tree planting. (a) Excavated soil deposited in such a way that the uppermost layer will again be on the top after refilling. (b) The stake must be installed before planting. (c) The formation of a flat mound in the hole before planting. (d) Roots spread out on the top of the mound with no bending. (e) On refilling, the hole's edges are broken in order to loosen a larger area of soil (no "potting"). (f) The soil is formed into a flat trough; irrigation water cannot flow off.

the cuts point downwards. Fibrous roots are to be treated carefully. Trees with bare roots are put up on a cone-shaped mound (Fig. 90) which is made on the bottom of the planting hole and upon which the roots are spread. No root should be bent to save space. Tree stakes must be set tightly into the firm ground before the planting, so that no roots will be damaged. A board is placed across the hole to ensure that the tree is not planted too deep. The root collar must be on the same level as the upper edge of the board because the loose soil around the roots has yet to settle. Any root loss should be compensated for by a corresponding pruning of the branches. The planting of trees with a soil ball wrapped in hessian (jute, hanf) proceeds in the same way. After the tree is put into the planting hole the hessian is loosened around the root collar but not removed completely. A stake is set in a slanting position so as not to damage the root ball. Generally, pruning of trees with soil balls is not necessary.

Planting of street trees (Plates 112—116; Figs. 91 and 92)

A soil analysis prior to planting will always be useful because it affords familiarity with local conditions. In fact the soil of most Central European towns, for instance, contains an excess of lime, the result of the ruins of centuries. Soil that does not meet the basic requirements should be changed completely wherever possible. This may, of course, have the effect of slowing down growth as soon as the tree roots grow out of the hole into the adjoining ground.

Plate 112. Street tree-planting in a raised bed with build-in ventilating and irrigating systems, chips and special stones (Photo Pessler).

The distance between buildings and curbstones is decisive for street-tree planting. If it is less than 5 m and the sidewalk is not at least 4 m wide, street trees should not be planted. The distance from tree to tree depends on the species and on other circumstances, but should not be less than 10 m (Fig. 29). For minimum distances between trees and public utilities see Table 63. All cut roots of existing trees must get a root curtain (page 236).

The planting hole for new trees should measure at least 2 × 2 × 1 m, but preferably 3 × 3 × 1 m. Across the root area, at least 20 m² of ground should

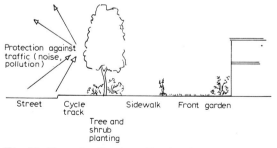

Fig. 91. Tree planting in a city street.

246

Plate 113. As Plate 112, with protective grating for the tree (Photo Pessler).

be kept open. If this area is not to be planted, it should be covered with slit flagstones or holed concrete slabs to let the air in. The slits or holes should be cleaned regularly. The flagstones are laid into 5—10 cm of grit or sand. Buffer stones and steel bows protect the trees against vehicles. The planting hole must be secured, technically and biologically, against the stress of the

Fig. 92. Diagram of a street in Brussels.

Plate 114. Staking trees (Photo Pessler).

TABLE 63

Minimum planting distances of street trees

From curbstones to centre of tree	1 m
From poles of overhead wires	4 m
From cantilevers of street lights of more than 7 m height	10 m
From sewers	2 m
From rainwater gratings	2 m
From conduits up to NW 400 (gas and water)	2 m
From pipelines above NW 400 (gas and water), high-pressure and long-distance heating	3 m
From underground cables (high and low voltage and telephone)	2 m

248

Plate 115. Staking trees on a town street (Photo Bernatzky).

town climate; shortage of water, air and nutrients. For warding off melting snow- and ice-water which contains salt, the whole surface area under the trees must be elevated about 20—30 cm above the adjoining ground with border stones (Fig. 91). If there is more than one tree, the entire border should be done in this way. At the time of planting a ring of drain pipes and irrigation pipes must be installed (plastic tubes, diameter 70 mm) in the root area. Vertical tubes, leading to the surface, must be covered with perforated caps. The installation depth of the pipes varies with the different species but will generally be 30—50 cm.

Plate 116. The stakes are too long (Photo Bernatzky).

Tree stakes

The bigger the planted tree the better must be the protection against the force of the wind which will tear off the new roots. Next to the vertical pole, triangular combinations and exchangeable quadruple steel pipe supports are recommended. Guy wires and steel cables should not be used in streets because pedestrians may stumble over them; they may be used in parks, however. Harris (1974) weighs the advantages and disadvantages of tree staking in the landscape, but these thoughts can also be applied to street trees. "An unstaked tree will have these characteristics: greater caliper at the trunk's base, less growth in height, greater trunk taper, less wind resistance, subject to less stress per unit of trunk cross-sectional area at the support point, more uniform xylem tissue (wood) for supporting itself upright, no rubbing and girdling injuries and a larger root system. There are three reasons for supporting and protecting the tree with staking: protecting the trunk from damage by equipment, vehicles and vandals, anchoring the root system, and supporting the trunk in an upright position."

Watering

In the process of transplanting, the water balance is considerably disturbed. It is not restored with the planting of the tree (Kozlowski, 1971b).

250

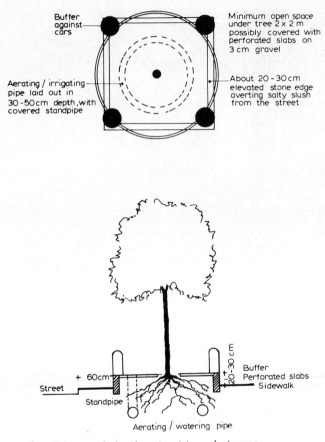

Fig. 93. Scheme of planting street trees in towns.

The most important cause of the death or slower growth of transplanted trees is desiccation as a result of transpiration. It can be reduced by judicious pruning or by the use of antitranspirants.

Even well-rooted trees, however, will not rapidly restore their water balance. The ability to withstand desiccation depends on the ability to reduce transpirational water loss, to absorb water at high rates or to endure dehydration. Davies et al. (1972) investigated these relations. They showed that the transpiration capacity of trees varies between species and clones within the species. Leaf area, root/shoot ratio, structure and size of stomata, stomatal frequency, control of stomatal aperture and leaf anatomy are the most influential factors in transpiration (Table 64). To a large degree the capacity to tolerate water loss depends on the diffusive resistance of the leaf, which can be changed by antitranspirants (antidesiccants) (Table 65). There are two types of antitranspirants: metabolic inhibitors and film leaf coatings

TABLE 64

Stomatal length and distribution of 28 species of trees found on the University of Wisconsin Campus (after Davies et al., 1972)

Species	Stomatal length μ	Stomatal frequency/mm^2
Acer saccharinum	17.29 ± 0.25	418.75 ± 12.56
Acer saccharum	19.28 ± 0.5	463.39 ± 18.94
Acer negundo	21.59 ± 0.32	233.93 ± 13.98
Betula nigra	39.36 ± 0.60	281.25 ± 11.38
Betula papyrifera	33.22 ± 0.56	172.32 ± 10.49
Catalpa bignoniodes	23.21 ± 0.71	328.57 ± 13.23
Crataegus sp. I.	22.27 ± 0.36	399.11 ± 15.49
Crataegus sp. II.	37.43 ± 0.61	221.43 ± 11.64
Fraxinus americana	24.84 ± 0.25	257.14 ± 14.59
Fraxinus pennsylvanica	29.33 ± 0.65	161.60 ± 15.82
Ginkgo biloba	56.30 ± 0.89	102.68 ± 6.83
Gleditsia triacanthos	36.10 ± 0.79	156.25 ± 4.24
Hamamelis mollis	25.27 ± 0.39	161.61 ± 6.79
Juglans nigra	25.68 ± 0.41	341.96 ± 5.15
Malus sp.	23.84 ± 0.25	219.54 ± 12.32
Populus deltoides	30.36 ± 0.19	163.39 ± 6.00
Prunus serotina	30.45 ± 0.32	306.25 ± 8.81
Prunus virginiana	27.08 ± 0.41	244.64 ± 9.96
Quercus rubra	26.71 ± 0.61	532.14 ± 11.14
Quercus macrocarpa	23.99 ± 0.29	575.86 ± 14.58
Quercus palustris	30.86 ± 0.34	530.36 ± 18.24
Rhus typhina	19.38 ± 0.19	633.93 ± 14.56
Robinia pseudocacia	17.63 ± 0.32	282.14 ± 11.36
Salix fragilis	25.52 ± 0.39	215.18 ± 8.63
Syringa sp.	29.10 ± 0.49	324.11 ± 9.77
Tilia americana	27.21 ± 0.33	278.75 ± 10.07
Ulmus americana	26.27 ± 0.28	440.18 ± 18.56
Vitis vinifera	29.73 ± 0.23	120.48 ± 6.21

(Smith, 1971). Davies et al. (1972) found that overall recommendations are hard to give because of variations in the rate of transpiration between and within individual species. Overdoses can cause injuries: formation of lesions on leaves, chlorosis, abscission and leaf mortality. Film-type antitranspirants control water release better in species with small stomata, but they block not only the outward diffusion of water but also the inward diffusion of carbon dioxide and, thereby, photosynthesis. All antitranspirants raise leaf temperatures (page 45). Kozlowski (1971, cited in Davies et al., 1972) recommends irrigation about once a week to a depth of 15 ins. in well-drained soils, or somewhat less often in tight soils. Smith (1971) recommends application of antidesiccants for conifers and other evergreens in winter to reduce transpiration and to prevent drying out.

TABLE 65

Effect of antitranspirants applied in various dosages on stomatal resistance of well-watered white ash plants (after Davies et al., 1972) *

Antitranspirant	Concentration (%)	Stomatal resistance (sec/cm)	Concentration (%)	Stomatal resistance (sec/cm)	Concentration (%)	Stomatal resistance (sec/cm)	Concentration (%)	Stomatal resistance (sec/cm)
TAG	100	32.0	50	19.2	20	12.4	10	19.8
CS 6432	10	15.0	5	16.3	2.5	8.5	1	5.6
Folicote	20	20.2	10	18.5	5	18.5	1	7.2
Keykote	20	25.7	10	5.9	5	8.2	1	13.0
Clear Spray	100	17.9	50	16.3	33	15.9	20	15.9
Wilt Pruf	50	6.2	33	10.4	20	10.4	10	—
Aqua Gro	20	39.4	10	14.9	5	—	1	5.5
Vapor Gard	20	16.6	10	12.7	5	12.4	1	4.6

* Stomatal resistance of controls = 4.3 sec/cm.

8.2. *Transplanting large trees*

The bigger the trees which have to be transplanted, the greater is the interference with their living requirements. Trees in the open country have a wide-spread root system. The older and bigger the tree, the further away the fibrous roots are from the trunk. The same applies to trees which are to be transplanted from one garden into another. However, these trees can be prepared for transplantation.

Round the tree a wide trench (30—50 cm, depending on the root system of the tree) is dug, the radius of which should be at least five times the diameter of the trunk at a height of 30 cm (Fig. 94). In the Spring of the first year half of this trench is dug; not in a semicircle, but in short sectors all round the tree. The roots are carefully exposed and cut as specified for installing a root curtain (page 236) but without incorporating wire netting. Then the trench is filled up again with a mixture of compost, ground peat and tree food. The following Spring it is the turn of the other sectors. When the tree is dug up another year later, it has formed so many fibrous roots in the meantime that it easily takes root in the new location. Spring and Autumn are the best times for planting trees treated in this way.

Large trees are often transplanted in the evening or at night because the microorganisms which promote rapid root growth following transplantation are often damaged or even killed by the sun's radiation (Plate 117). There

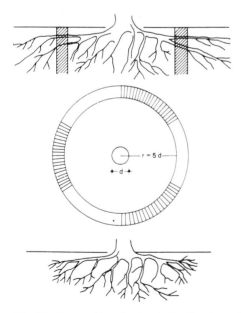

Fig. 94. Preparation for transplanting large trees.

Plate 117. Birch and maple planted fully green in June 1975 (diameter 50 and 50 cm) (Photo Copijn).

are various methods of transplanting large trees, from manual work only to fully mechanized transplanting (Goeke, 1972; Matthews, 1974). First of all the root ball has to be dug up. Its size depends on the tree species and on root extent, the age of the tree, its health and kind and the distance of the transport. Generally the ball is dug conically cutting the roots accordingly, sometimes with a steel wire of a power winch. During this work the surface of the root ball should not be stepped on. The dug-out ball can be secured in various ways:

by freezing in frosty weather after water has been poured over it; a method used only for trees with insensitive roots, wrapping up with burlap or "hessian"; a method used mostly with smaller trees.

tying up with wire netting,

attaching a chain basket, which is also used to hang up the ball for transportation.

tying up with ropes or cables,

constructing a container with wooden boards and sheets of steel, held together by ropes or chains,

constructing a wooden box (used in light sandy soils which will not hold a root ball),

building in the ball with concrete (very expensive and complicated).

For transport, the trunk must be especially protected against crushing of bark and cambium at all those points where the transporting implements are put on. This is done by wrapping with thick cocofibre ropes. The trunk may also be drilled through at breast height (diameter 45 mm), a thick-walled steel pipe pushed through the hole, and at its ends the ropes of the block and tackle attached for transport. Later on the drilled hole is blocked with a plug and treated with wound dressing. The terminal shoot of the tree can be protected by tying a pole onto it.

Various vehicles have been developed for tree transportation, from the simple roll to the crane truck and finally to the special transplanting truck or Tree Mover (Plates 118 and 119). The Tree Mover is a single machine which can carry out the entire operation of digging out the new planting hole, getting the tree to be transplanted from its old position, transporting it to the new one and planting it there. The digging out of the old root ball is done by hydraulic spades or cutting chains similar to a power saw. Depending on the size of the implements, the dimension of the soil ball will be 2 m or more in diameter. Trees up to a breast height diameter of 8 inches (= 20 cm) can be transplanted in this way. If a tree is too large for the machinery and has far-extending roots these will often be torn and crushed, thus causing decay (Plate 120). Because the tree is put into a planting hole dug out by the same machinery, stagnation of root growth is possible. Tree Movers are not suitable for sandy soils or rocky ground.

8.2.1. Aftercare following transplanting

In the new location the bottom of the planting hole as well as its side walls should be broken up to stimulate new root growth, particularly in compressed, compacted ground. Trees should not be "potted". After putting the tree into the new hole, all ball-securing devices (plastic material, etc.) must be removed except the wire netting and burlap or hessian wrapping. These must be opened up only at the root collar, and should stay in the soil where they will rot in time. An all-round covering of the root ball with sand helps to prevent decay and stimulates new root growth. The filling of the hole must be done carefully, that is, no earth-free cavities should remain any-

Plate 118. Transplanting of an old tree (Photo Copijn).

where. For the preservation of moisture in the transplanted root ball, belt tiles are put in vertically at the edges or irrigation wicks of spin fleece are used. The reduction of transpiration is important. This is achieved by the following measures:

Plate 119. Tree mover (Photo Bernatzky).

Plate 120. Tree roots damaged by transplanting with tree mover (Photo Bernatzky).

258

Plate 121. Protection of the trunk (Photo Copijn).

thinning of the crown (not just lopping the branches),
use of antitranspirants (pages 45 and 250),
construction of burlap sun sails (against the sun's radiation and reflection from buildings),
installation of water spray nozzles in the crown (to be used with care on ground which easily becomes muddy),
application of bandages around the trunk and the largest branches, consisting of loam and burlap, ropes of straw or reed mats, against drying out of the bark, sunburn, frost damage or gnawing by game (Plate 121).

All injuries the trees have suffered during transplantation must be treated at once and with great care.

8.3. Anchoring of trees

In a park and in the open landscape, large transplanted trees are supported by the use of wires or steel wire cables, stretching from the onset of the crown at a slanting angle to poles or anchorage points in the soil. They get tightened by application of cross bars and turnbuckles. The bark of the tree must be carefully protected by rubber padding. It is also possible to insert poles firmly into the open planting hole and tie the ball onto them with cables. This anchoring is invisible.

8.4. Inspection and estimation of trees

The inspection of trees in the city is of great importance. The reason for this is the increasing deterioration of their living conditions and the growing interference with their stability by traffic and public services. Street trees are especially endangered. They on their part endanger public safety because branches or whole limbs may break off, or the tree may collapse altogether. Therefore they have to be inspected regularly so that damage to people and property may be prevented in time. This can be done on an extensive scale by remote sensing (page 273). Just as important is the inspection and estimation of trees in a local area. Theoretically all trees should be inspected, from their planting onwards, over longer or shorter intervals. In the first years after planting this is necessary to evaluate their growth; later, from about the 25th year of their lives, to detect or prevent damage.

The authorities responsible for the safety of trees should enforce rules and regulations which unambiguously stipulate:
who carries out the inspection,
when and how it will be carried out (e.g., twice a year, before and after the shedding of leaves),
what is to be inspected,
how the inspection will be analysed.
In large towns the park departments should be responsible for this. In other towns special street divisions should be installed and adequately financed. The positions must be occupied by independent experts. In Table 66 an outline is given for the diagnosis of tree troubles and tree stability. The result of these investigations is laid down in an inspection book (Table 67) and/or an inspection card (Fig. 95). These notes can be used as valuable evidence at judicial hearings.

The examination of forks and trunks deserves special consideration. At a visible cavity inspection is simple. Otherwise the whole trunk has to be tapped carefully with a plastic hammer (with an iron core). If the wood gives a light sound it is probably healthy, but muffled or hollow sounds indicate that the heartwood may be decaying. In this case the trunk should be drilled through with an auger to check that the wood is still healthy all through. The extracted core shows the condition of the wood. Drill openings should always be closed with a plug. If a cavity in the trunk is discovered in this way it can be opened up carefully with a power saw and treated, but only a strong suspicion of decay justifies drilling. Tapping the trunk is simple and practicable. Other methods of investigation have not been successful (Ruge, 1971b):
X-ray photographs. They are expensive. It is impossible to distinguish

TABLE 66

Outline for the diagnosis of tree trouble *

1. General view

    11 Deviation from the characteristic structure of the particular species
    12 Unnatural discolouring of the foliage
    13 Premature fall of leaves
    14 Top dryness
    15 Pest and vermin injuries
    16 Anomalous growth

2. Position

    21 Tree too close to the road
    22 Crown hanging too deep into the road
    23 Crown extending into the roadway
    24 Tree too close to buildings
    25 Tree hemmed in between other trees
    26 Tree hemmed in between buildings
    27 Damage from construction work

3. Root areas

    31 Overfilling with earth, concrete, asphalt, etc.
    32 Lawn covering
    33 Other compactions
    34 Soil removal
    35 Root losses caused by sewers, pipe lines, etc.
    36 Soil contamination by gas, oil, de-icing salt
    37 Unevenness of the surface (tree roots, loose flagstones, etc.)

4. The trunk

    41 Damage of the root collar
    42 Decay at the root collar
    43 Bark damage (caused by dogs, heat reflection)
    44 Other bark damage (frost cracks, lightning)
    45 Bark injuries by vehicles
    46 Wet spots on the trunk
    47 Cavities in the closed trunk (invisible)
    48 Open cavities or centres of decay

5. Onset of the crown

    51 Narrow, pointed forks
    52 Slit up forks
    53 Decaying forks
    54 Broken off branches

6. The crown

    61 Breaking out of branches
    62 Decaying holes
    63 Dead and broken branches
    64 Extending branches
    65 Top dryness
    66 Bunch of branches as result of lopping

TABLE 66 (continued)

Corrective measures

22. Position

      221 Removal of trees
      222 Removal of branches
      223 Cutting back, cabling
      224 Thinning out
      225 Current observation

32. Root area

      321 Removal of the overfilling
      322 Removal of lawn, replacement by pavement
      323 Aeration, installation of irrigation system
      324 Fertilizing
      325 Root curtain
      326 Soil analysis in a laboratory; aeration and watering
      327 Removing of unevenness
      328 Planting of low shrubbery

42. The trunk

      421 Careful digging up
      422 Treatment of decaying spots and other damage
      423 Wound treatment, also prophylactic measures: Lacbalsam, tree basket, special cuffs
      424 Nailing of bark, screw rods, wound treatment
      425 Wound treatment, setting up of buffer stones and steel bows
      426 Opening up of the tree and treatment
      427 Tapping off the tree, opening if needed
      428 Cavity treatment

52. Onset of the crown

      521 Draining of the fork
      522 Screw rods
      523 Treatment of decaying spots
      524 Wound treatment

62. The crown

      621 Wound treatment
      622 Cavity treatment
      623 Removal of branches
      624 Cutting back of branches
      625 Ascertaining reasons, pruning
      626 Thinning out

* These numbers have to be registered in the inspection card Fig. 95.

TABLE 67

Inspection book for trees

Street: ....

Date	House no.	Species of tree	Detected trouble	Treatment	Stability yes/no
			Please give numbers		

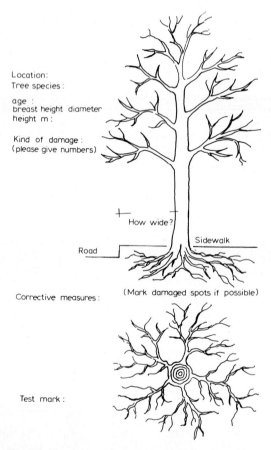

Location:
Tree species:

age :
breast height diameter
height m :

Kind of damage :
(please give numbers)

How wide?

Sidewalk

Road

(Mark damaged spots if possible)

Corrective measures:

Test mark :

Fig. 95. Example of a card used in tree inspections.

between fully healthy wood and the air taking the place of fully destroyed wood. Besides this, technical requirements and X-ray protection present problems.

Radioactive rays. As above.

Ultrasonic waves. The bark slows down the speed of the waves so much that the whole bark would have to be removed. For trees with diameter above 40 cm this method fails totally.

Puncturing method according to Gillwald. The steel needle allows an insertion of only 10 cm.

Increment drill. Its application weakens the tree too much in the long run.

Tractive tests by rope or tractor must be avoided in all cases because they will — without exception — cause serious damage.

However Tattar (1975) reports that "electrical measurements enable us to detect infections of pathogens before the tree reveals any visual symptoms of the disease. Change in membrane permeability is one of the first events that occurs in many diseases. Loss of selective permeability and subsequent electrolyte loss by tissues around the cambium will affect the electrical conductivity of the fluids in the transpiration stream. Increase in electrical conductivity in the cambial zone may serve as an indicator of early pathogen attack. Initial growth of pathogens in the conductive elements is thought to impede the transpiration stream. The movement of the sap to the leaves results in streaming potentials and these biopotentials will change if the rate of flow is changed. Impedence of vessels by fungi will change the biopotentials and may indicate an early infection of a number of plants". Excavations in the root zone and adjoining area should be done only by hand and with the greatest care as otherwise serious damage will be caused.

9. TREE NUTRITION

Under natural conditions the circulation of mineral substances in the eco-system is largely closed. The mineral substances absorbed by the trees are returned to the soil in the form of organic matter; there they are mineralized again and absorbed by the roots once more, bit where no organic waste is left to the ecosystem, used-up nutrients must be replaced.

On fertilization in general, and the feeding of trees in particular, opinions differ very much (Himelick et al., 1965; Murphy and Meyer, 1969; Neely et al., 1970). An analytical test may clearly show the actual needs of the soil. Expenses for this will be compensated by economical fertilizing.

Those who advocate humus-farming suggest the use of powdered rock in preference to synthetic fertilizer to replace used-up minerals (Rusch, 1968). In this way the content of clay crystals in the soil will also be increased. The so-called nutrients are auxiliary materials of the vital functions between the soil and the plants growing on it. Therefore they should be replaced only by the original minerals which will be disintegrated and made available in the soil whenever the plants need them. An artificial dosing of minerals can never replace the natural exchange of minerals between soil and plant.

Especially for street trees living under stress conditions, the best manure is still a well-seasoned, matured humus. A spontaneous balance of ions predom-inates in such soil. A drop in the basic saturation rate or the pH-value does not occur and a wash-out of nutrients is prevented.

9.1. Fertilizing

Ruge (1972a) stated that trees in Western Germany, particularly street trees, suffer from a lack of K_2O, MgO, P_2O_5, NO_3, B and Mn while there is a large surplus of CaO and NaCl. This may also be true for cities in other coun-tries. Trees require an abundance of K_2O and P_2O_5 to further root growth and drought resistance. Mature trees no longer require as much nitrogen because this substance compels vegetative growth and lowers physiological resistance. Ruge suggests an NPKMg-ratio of 6/10/18/2 and 10/15/20/2 plus B and Mn (trace elements). Physiologically, the fertilizer should work like an acid and be free of Cl^- (damage by de-icing salts). Nitrogen should be given neither in the form of NH_4 nor as urea, because too much oxygen is required in the soil to disintegrate them. Great differences are noted concerning the formulation of the different nutrients. Beardall (1972) suggests an NPK-ratio of 10/6/4 whereas Pirone (1973) mentions one of 16/8/16 or 7/40/6. Generally an amount of 1/2 kg/cm bhd (breast height diameter) of the trunk, or 2 lb/inch bhd, is recommended. Lately Bettes has proposed (according to Pirone) another calculation: trunk diameter × trunk diameter divided by three (in lb). While with the former method a tree of 12 inches bhd will get twice as much fertilizer as one of 6 inches bhd, the new method allows the larger tree 4 times as much. This seems to be a very large amount. Moreover the quantity has to be adjusted according to the formulation measure and to the dimensions of the crown as well as to the kind of soil. In Western Germany a combined organic—mineral compound of tree manure (NPK 6/8/10) is widely used with great success ("Maurers Tree Food").

9.2. Ground feeding

The tree derives no benefit from fertilizer scattered on the lawn below it or just scraped superficially into the ground. Only the lawn or other ground cover shrubs will be fertilized in this way, never the tree.

The usual practice for feeding, especially with older trees, is the punch bar or auger application. Holes with a diameter of 5—10 cm are punched in the soil with a bar, or drilled with the auger, or washed out with water at a slight angle downwards (Plates 122 and 123). The drilling or washing out is better

Plate 122. Making holes for ground feeding in asphalt (Photo Pessler).

Plate 123. Making holes for ground feeding in soil with auger (Photo Pessler).

than punching because the latter compacts the soil. The depth of the holes depends on root growth, but will generally be somewhere between 30 and 60 cm. The holes are now filled with tree food up to about 10 cm below ground level. In open soil they can be trodden down afterwards; in paved surfaces the remaining 10 cm will be filled up with gravel (7—15 mm diameter) (Figs. 96 and 97). The feeding should cover the area of the crown's spread plus 2—3 m beyond. It should start at a distance of 75—100 cm from the trunk and proceed in circles. The holes on these circles should be 80 cm apart. In a narrower location, this distance may be reduced to 60 cm. There should be 4—8 holes per m². If the ground under the tree is covered with low

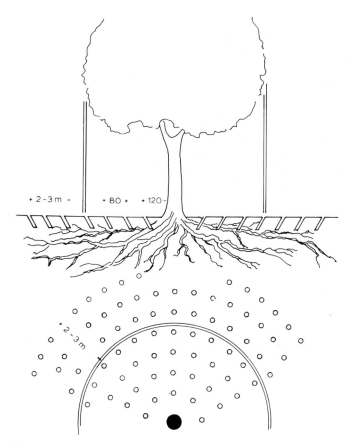

Fig. 96. Tree feeding (ground feeding).

shrubbery, some tree food is scattered there too. This method should be used to fertilize:

all trees where fallen leaves have been removed,

all trees in a lawn,

all trees on very poor soils and trees that are hungry for other reasons,

all trees with strong undergrowth,

all trees on concrete or asphalt pavements,

all trees to be transplanted (these should be fertilized one year before the transplant),

all freshly planted trees; fertilizer should be given outside of the new planting hole to induce the new roots to grow out of the planting hole "pot",

all trees treated surgically (ideally, fertilization precedes surgery by one year),

every single tree in a group of trees.

268

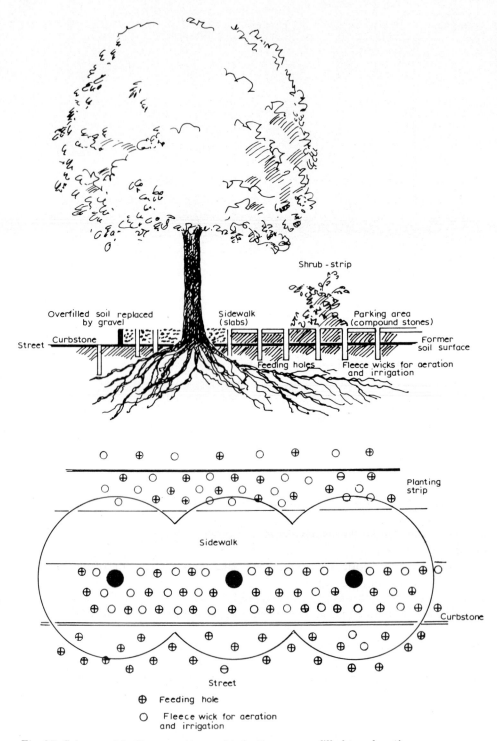

Fig. 97. Scheme of feeding, aerating and irrigating an overfilled tree location.

Plate 124. Feeding with needles (too close to the trunk!) (Photo Pessler).

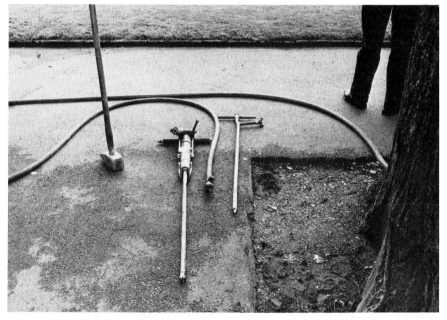

Plate 125. Tools for needle-feeding (Photo Pessler).

Mineral ferilizers can also be applied in liquid form by means of feeding needles (Plates 124 and 125) or similar devices. Dry fertilizer may also be put into holes made by compressed air drills. In the U.S.A., fertilizer packets containing soluble inorganic food are on the market (Attoe et al., 1970). They are put into spade-dug holes 20—25 cm deep within the root area and covered with soil. According to Attoe et al. (1970), the vapour pressure in the soil regulates the delivery of nutrients. With decreasing vapour pressure in the Autumn the delivery of fertilizer is stopped. Only when the soil warms up again in Spring does the vapour pressure increase again, releasing plant nutrients once more. Fertilizing with tree food spikes is time-saving. These are simply driven into the ground with a strong carpenter's hammer.

9.3. Foliar feeding

Foliar feeding is assuming increasing importance as a substitute for ground feeding where trees are growing in concrete, asphalt or paving. Foliar feeding works rapidly but its result is short-lived. It may be used in addition to ground feeding. Young leaves absorb foliar feeding better than to older ones, but on the whole the effect depends on the size of the leaf surface index (relation of leaf surface to ground surface). It can be carried out from the beginning of sprouting in Spring until early Autumn. Thereafter no feeding should take place as this would result in another growth impulse. The dosage has to be exactly as prescribed to avoid scorching (Bukovac and Witwer, 1957; Tukey, 1952).

9.4. Soil-improving agents

These are a particular kind of nutritive aid. They affect the tree indirectly, via the soil. Natural soil-improving agents have been well known for a long time: sand for clay soils; loam for sandy soils; peat, etc. A biological way of improving soils consists of a vaccination of symbionts of physiological bacteria. These are bacteria of such species as are found in the whole cycle of nature from man down to fertile humus (Symbioflor-Humus-Ferment). Spreading an organic soil cover (nutritive cover) of compost about 1 cm thick *onto* the soil after planting is also part of soil improvement. Digging organic material *into* the soil is dangerous, especially in poorly drained soils.

Lime. The effect of lime on the soil is well known especially with regard to the clay—humus complexes. Through a shift of the soil reaction towards the neutral zone, soil life will be activated and organic substances developed, which leads to an aggregation of soil particles.

Polysaccharides. These are of a polyuronic structure and are formed by microorganisms and higher plants. They serve as a binding agent between soil particles. Algae possess a great many such highly polymeric substances. The algae powders (e.g. Alginure) manufactured from them contain $CaCO_3$, polysaccharides, uronic and alginic acids. Application of these polysaccharides improves the soil's water-retaining capacity as well as the ventilation. Root growth will be furthered and the soil structure considerably improved.

Bycobact. This is a liquid concentrate with 60% organic components: lignin, amino acids, hemicelluloses and living soil bacteria. It contains N (4%), K (3%) and Ca (2%). It activates humus development and soil life, and it counteracts damage done by de-icing salts. It is applied in liquid form, diluted with water, by means of feeding needles or irrigation devices.

Polymeric linear colloids. Lately more and more synthetic improvers are offered (Schaller, 1975; Pruen, 1975). These include synthetically manufactured polymeric linear colloids (linear polymers). Successful use of these, however, depends on an abundance of clay colloids in the soil. They are supposed to further breaking up and crumbling of the soil, and to improve aeration and the water economy.

Liquid mulching. This is carried out by spraying an emulsion (bitumen, latex, "Weimulsion") onto the soil in a thin filmy layer. This is supposed to have the same effect as mulching with natural substrata. However, only the existing condition of the soil structure will be fixed in this way.

Alkaline silicates. These are finely dispersed amorphous colloids, Na- or K-salts of fluosilicic acid (e.g. Agrosil). They regulate the condition of the soil pores and develop soil colloids. They also set free bound phosphate and make it available to the plants. Agrosil is on the market in two different forms. One form is that of highly molecular, porous gels, slowly movable with surface activity; they are able to store water and nutrients and they narrow soil pores, buffer salts and net soil particles. The other form is a low molecular movable sol with chemical activity; it prevents phosphates from being fixed, binds heavy metal ions, stimulates root growth and also has a netting effect. Both forms improve the location soil and hence growing conditions. They also ensure better utilization of fertilizer dosages. The observed deepening of the roots makes accessible larger reserves of water and nutrients in the soil. Fairly great losses at fresh tree plantings can be avoided. Agrosil is not, however, a normal fertilizer. It can either be mixed into the soil at the bottom of the planting hole, or scattered onto the surface after planting, or sprayed on as a solution after being dissolved in water.

Synthetic foams with open cells (Hygromull). These are synthetic foaming resins of urea and formaldehyde. They possess a hollow volume of 70% and contain 30% nitrogen of which only a small part is available to plants (3—5%

yearly in a soil normally supplied with lime). The water-retaining capacity amounts to 50—70% so that water storage is high. Soil—physical effects of the foam could not be detected in a field test (Schaller, 1975). It should be brought into the soil thoroughly wetted because primary moistening of the material takes place very slowly. The output of water also proceeds gradually.

Synthetic foams with closed cells (Styromull, etc.). These cannot absorb any water but serve only for breaking up and loosening of dense soils. They are chemically neutral, biologically sterile, and they do not age or rot (neither in chemical nor biological decomposition). Hygropor is a mixed product of Hygromull and Styromull and is therefore capable of absorbing and storing water as well as of supporting soil aeration. It should be borne in mind that the three last mentioned materials are synthetics and not biological substrates. They will reduce the amount of biologically active soil in proportion to their quantity. They will compete with the living active forces of the soil. There may be reasons to use them in exceptional cases such as the lay-out of a roof garden or similar exposed sites.

10. THE USE OF FALSE-COLOUR, INFRARED PHOTOGRAPHS IN REMOTE SENSING TO CHECK THE HEALTH OF TREE POPULATIONS

The importance of aerial photographs in the registration of the landscape, its different forms and outcrops of vegetation, has long been known. Black and white films were used initially, coloured films later. With both types the differentiation obtained was not enough to make possible an exact interpretation of the reproduced green areas. The expanded use of coloured film resulted in the use of so-called false-colour infrared (IR) film. This is sensitive to the near infrared part of the spectrum (wavelength 700—900 nm), which is invisible to the human eye.

The structure of the IR film is fundamentally the same as that of normal colour film, except that spectral sensitivity is widened towards the near infrared and changed in reproduction. It ranges between blue, green, red and infrared. In order to remove the blue, a blue-absorbing filter must be added. Due to the change of sensitivity towards higher spectral zones this film will, when developed, show the following results in the positive:
green becomes blue,
red becomes green,
infrared becomes red (Fig. 98).
The reason for the choice of this method, which extends beyond the visible light into the infrared zones of the spectrum, is as follows. A normal colour photograph will show the green of photographed trees in various tints and shades, differing from each other but not clearly distinguishable. As plants absorb about 80—95% of light at a wavelength of ca. 400—700 nm (visible to the human eye) in photosynthesis, only 5—20% will be reflected. Of the invisible infrared (700—900 nm) trees and other plants absorb only 20% and reflect 80%; this intense reflection is used in the IR picture. It is the cell structure of the leaves, especially the mesophyll and spongy parenchyma, which causes this reflection of the near infrared light. Conifers have a different cell structure, so that their reflection amounts to only about 40%; on the infrared picture they appear darker than deciduous trees. When healthy plants are photographed with infrared film, the upper layer of the film receives most of the light and the trees appear typically red. Beyond the upper layer lie the green-sensitive and the red-sensitive layers; the last one, being the lowest, receives very little light when healthy plants are photographed. Damaged plants lose to a considerable extent their ability to reflect the near infrared, so that the medium and lower layers of the film will receive more light. The colours change accordingly. Dead trees or dead parts

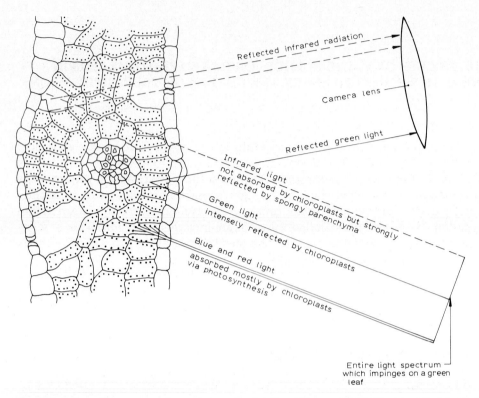

Fig. 98. Schematic cross-section of an oatleaf (after Eames and McDaniels; cited in Stellingwerf, 1973).

Reproduction of false colour film

Reflection	Negative	Positive
Green	Yellow	Blue
Red	Magenta	Green
Infrared	Cyan	Red
Green + red	Red	Cyan
Red + infrared	Blue	Yellow

of plants will appear bluish-green (cyan) in the positive (Pollanschuetz, 1968). Normally, IR pictures are taken as positive. Their exposure range is larger than that of pictures on a negative film and printing flaws are eliminated.

The scheme of colour reproduction of the IR film given above is modified by a series of factors. For instance, the natural colour of the leaves plays a part. Lighter coloured leaves also appear lighter on the IR picture. Oaks are darker than beeches, firs are darker than deciduous trees. The Autumn colouring shows up in yellowish and whitish tints. White flowering trees, like

Sophora and *Robinia*, have the appearance of damaged trees. Trees that lack water and have flaccid leaves may be moved more by the wind, so that the undersurface of the leaves will show up in the aerial picture. Leaves having light or hairy undersurfaces will change the reflected picture accordingly. At the same time, the shade pattern of the flaccid tree top is changed (Kadro and Kenneweg, 1973).

10.1. What will be observed on the IR picture?

Generally speaking it can be ascertained whether trees and other vegetation areas (including lawns) are healthy or not. If not, the colour gradations show the different grades of damage. Of course, only the fact of damage will be apparent; the aerial picture will not say anything about causes. The latter (and there are mostly more than one) must be discovered by inspection from the ground and by special tests (botanical, physiological, chemical and soil-scientific) and by a study of overall growing conditions. Therefore, on the basis of aerial pictures and the tests made afterwards, an interpretation key must be set up which will be valid only for the pictures of one flight, as the colour reproduction and quality of the pictures depend on a variety of different factors (film material, weather, exposure, etc.) which may alter with each flight.

A point of controversy in the use of these photographs is the question of whether "invisible" damage can be recognized (early perception). Some authors suggest that this is the case, but it is a fact that visible symptoms of damage, although they may be inconspicuous or hidden, can be seen better on the IR picture. This is, of course, something totally different from the perception of symptoms that are completely invisible. It must also be remembered that necroses of leaves in the lower part of the tree top cannot be seen on an aerial photograph, just as those in the upper part of the tree cannot be seen in an inspection from the ground. Dead trees, bluish-green on the picture, can be distinguished from an asphalt surface only by the shadows they cast.

Taking into account the above details, false-colour IR pictures can be quite helpful for the recognition of damage caused by
de-icing salts spread in winter,
restricting the root area by paving with asphalt or other road building processes,
leakage of gas pipes,
soil compaction and poisonous remains (oil) in the soil,
shortage of water,
lack of nutrients,
enrichment of the soil air by SO_2 or other toxic material,
overheating of the tree tops during summer time by intensely reflected radiation,

injury to the roots by digging; injury to the trunk by vehicles,
biotic damage (e.g. by insects or fungi) (Kadro and Kenneweg, 1973).
At Mannheim (Western Germany) it was found that injuries to the roots
caused by digging were noticeable on the IR pictures before they could be
recognized from the ground (Wawrik, 1974, unpublished report of the
Garden Department of Mannheim).

10.2. Advantages of false-colour IR photographs

(1) Red-looking trees as seen on infrared photographs are more striking
to the observer than green trees on normally coloured photographs.
(2) The IR picture, with its greater range of tints and shades, makes it possible to recognize damage more quickly and easily than is the case with, for
instance, a black and white panchromatic photograph, or a regular colour
film. Treatment may then begin sooner and will be less troublesome.
(3) The aerial picture also makes it possible to photograph trees on private
grounds which are not easily accessible, especially those behind buildings.
(4) The IR pictures are more objective, eliminating many subjective possibilities of failure which frequently occur with inspections from the ground.
(5) From the air, all damage will be shown synchronously. Normal inspections from the ground take an extraordinary amount of time, during which
the conditions of the tree may alter.
(6) The condition of the trees on the day of the flight will be permanently
established. Doubts and errors of interpretation can always be investigated
later from the photographic material.
(7) Synchronous photography of wide areas will speed up landscape mapping, which will also be more complete.
(8) Interpretation does not have to be done during the flight, but can be carried out when more time is available for it.
(9) Interpretation may be carried out in an office or laboratory where workers are not likely to be distracted, as may be the case in the open air.
(10) These photographs are particularly important in providing evidence of
damage to plantings and green areas, resulting as an after-effect of building
activities, construction of skyscrapers and their foundations, construction of
wells, road buildings, etc. In all cases pictures should be taken prior to the
beginning of such construction work and after it, in order to allow comparisons to be made.

10.3. When to take aerial pictures

July and August are supposedly the best times of the year to take aerial
photographs. At any earlier time only the most severe damage will show up.
Later, the beginning of the Autumn colouring changes the degree of reflection and makes exact interpretations impossible. Noon is the best time of the

day to avoid disturbing shadows. The best colour contrasts will be achieved with a slightly overcast sky (high veils of clouds). Very cloudy weather will impair the quality of the pictures. If they are taken on different days, differences in colour reproduction may occur and interpretation will be doubtful. The most favourable scale is 1 : 5000. The use of larger scales is more expensive and makes interpretation more difficult on account of the larger number of pictures. Smaller scales yield vague results. To minimize distortion, cameras with a focal length of 30 cm are to be preferred. When these are used, the trees in the margins of the pictures will not be covered up as much by high buildings or other trees. Inquiries may be addressed to any companies conducting such aerial flights.

10.4. Interpretation of the photographs

The interpretation of infrared aerial photographs is relatively easy to learn but, by comparison with the present stock of trees (tree register), the individual species should be known since, as already mentioned, trees have their own characteristics on an aerial picture. It is advisable that prospective interpreters become acquainted with the appearance of damaged trees in the field and compare them with the reproduction obtained on the IR picture. For good interpretation an illuminated table is required with a high-intensity bulb or, alternatively, a good magnifying glass or a stereoscope (if the pictures overlap sufficiently), possibly with a binocular telescopic magnifier. For practical reasons the photographs are put into foliage bags which can be clearly marked. The better the interpreter knows local circumstances, the species of plants and their growth symptoms as well as the possible causes of damage, the better the interpretation will be. During interpretation and the resulting record filing of the pictures, care should be taken not to expose photographs to the light more than is absolutely necessary.

10.5. Drawing up a tree register

The extent of damage to the trees as shown on the photographs will be copied in the form of sketches, which need not be true to scale but which do in some way show, even if only by means of a simple line, the streets concerned and their trees. On these sketches the trees will be marked, with a blank circle or a filled one (Fig. 99), including the intermediate values 1/2, 1/4, 3/4 circle, depending on the extent of their damage. Kardo and Kenneweg recommend the following markings:

0 = Healthy trees; without interpretable leaf discolouration, necroses or other signs of damage. (On the aerial picture: red.)

1 = Slight to moderate damage; necroses, leaf discolouration distinguishable but healthy foliage predominant.

278

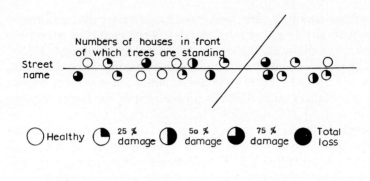

Fig. 99. Tree register.

2 = Serious damage; distinct necroses of the leaves and heavy losses of foliage characterize the appearance, there is little healthy foliage left. (On the aerial picture: greyish, brownish or greenish-white shades dominate over red.)

3 = Dying or dead trees; in July they are completely or almost completely without foliage. (On the aerial picture, bluish-green.)

The observed damage may also be drawn on a regular scaled town plan. If each level of damage is differentiated by a special colour or a certain length of line, a complete view will be obtained at once. The ascertained damage can be written on the cards of the tree register together with other details such as the quantity of thawing salt used, and the type of ground the trees are standing on (asphalt, lawn, etc.).

10.6. Costs of picture flights

The costs of such flights are not small and depend on the size of the area to be covered and on the selected scale. A considerable reduction in costs may be achieved if more than one administrative department is involved. For other purposes the IR pictures may be used also as black and white or coloured copies:

(1) For the planning of new green areas, of towns, landscapes and of whole regions;
(2) Especially for traffic planning (counting, preparation of new lines, drawing up an inventory of the trees on it and all other obligatory points);
(3) Surveyors' work;
(4) Reconstruction.

10.7. Infrared thermography

Infrared sensitive films can be used to measure the soil surface temperature of areas with and without vegetation. This measurement is based on the Stefan Boltzmann law, according to which the radiation energy of an ideal body (black) depends on its own temperature. These measurements are taken in the wavelength range of 5—13 microns or, to be more accurate, in the "windows" 2—5 and 8—13 microns. In the first range short-wave distur-

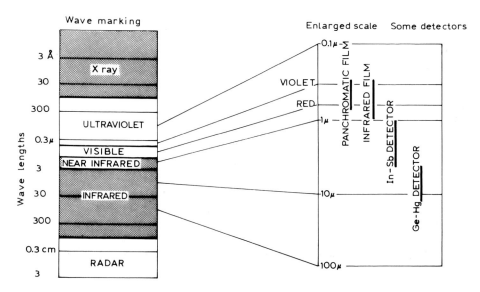

Fig. 100. Classification of the electromagnetic spectrum with some detectors (after Stellingwerf, 1973).
1 m = 10^3 mm = 10^4 micron = 10^9 nanometer (nm)
1 μ (micron) = 1/1000 mm
1 nm = 1/1000 μ

Wavelengths:	violet	400—440 nm
	blue	440—495 nm
	green	495—580 nm
	yellow + orange	580—640 nm
	red	640—750 nm

bances, reflected on the ground, may occur in the daytime, in addition to further disturbances caused by absorption bands of steam and carbon dioxide. Therefore it is better to use the range of 8—13 microns, when measurements are made during the day; at night the selection of wavelength is not decisive (Voelger, 1972).

The radiation is measured by means of half conductor detectors. With them, the area to be covered is scanned point by point in lines (Fig. 100). The pictures taken with the scanner are similar to regular aerial photographs, but the visible differences in light and dark areas are caused mainly by the differences in ground surface temperature. Coloured "Aequidensiten" pictures (equal density) may be prepared from the photos and on them the differences of the continuous grey scale appear as colour shades of different densities. The scale should be read from red (warm) via yellow and green to blue (cold). It must be remembered that the photographs obtained show only the ground surface temperature and not the actual air temperature, although the two are to some extent related. These thermographic pictures show relationships between a kind of vegetation and its form, and between the origin of cold air and its movement; they therefore provide information about the exchange of air and ventilation. This aids the town planner and landscape designer in the exact planning of ventilation and fresh air supply. However, in order closely to observe the origin of cold air and its movements, photographs must be taken at short intervals, in rapid succession.

10.8. Summary

False-colour infrared pictures are a valuable aid in the recognition of damage to trees and green areas. The causes of damage cannot, however, be ascertained from the air. Therefore an inspection from the ground and laboratory work are also necessary.

Infrared thermographic pictures show the influence of vegetation on air temperatures near the ground and so indicate how atmospheric conditions may be improved by securing a good supply of fresh air from vegetation areas (green plains, meadows, fields, woods). For this reason they are a helpful aid in the planning of atmospheric ventilation and fresh air supply for cities and overcrowded regions. (Further references: Hubbard and Grimes, 1972; Lundholm, 1972; Wert, 1972; Dyring, 1973; Genderen, 1974; Harney et al., 1973; Schuerholz and Larsson, 1973; Schneider, 1974.)

11. CONTROL OF PESTS AND DISEASES

This may take the form of preventive measures (hygiene) or direct control (therapy) (Butin and Zycha, 1973; Schwerdtfeger, 1970).

Hygiene. This consists of satisfying all ecological and location requirements such as choice of the right species for the location, cultivation and planting measures including fertilization, furthering resistance against pathogenic influences and protecting useful organisms.

Therapy. A correct diagnosis is essential for taking adequate measures for direct pest control. This includes, above all, the prognosis whether or not the infestation will take on abnormal dimensions since only when this is the case are severe cuts into the biocenosis justifiable. Control measures may be mechanical (like the gathering or catching of insects and removal of infected parts) or chemical or biological.

11.1. Chemical pest control

This is widespread today (Beroza, 1970). It works using toxic substances which kill pests through disturbance of physiological processes. Obstructive substances such as antibiotics will restrain the development of infecting agents. Repulsive substances affect the sensual organs of the pests, e.g. spreading bad-smelling products. Tempting substances will entice the insects so that they can be destroyed in large quantities. Sexual baits belong to this group. Poisons may be peroral (taken in through the mouth), percutaneous (through the cuticle) or systemic (through the sap flow). They can be oligotoxic or polytoxic. Occasionally they can also be phytotoxic and will thus damage the host of the infecting agent too.

The collective term for all substances used for killing noxious microorganisms and animals is pesticides. These include, amongst others, the fungicides, insecticides, rodenticides and nematocides. They are produced from organic as well as inorganic matter. Their number is immense and a special science concerns itself with them. The effect of a so-called timeless poison will not heighten when the quantity of poison is increased; instead, it will heighten rapidly at first, then slow down and finally there will be hardly any effect at all. Therefore quantities larger than those labelled produce no better effect. The period of effectiveness is of no consequence. This is not the case with time-bound poisons. Minimum doses of these may be successful if the period of effectiveness is long enough (e.g. gas). The susceptibility of pests to poisons varies according to species, genus, stage of development, age,

TABLE 68a

Insect and disease control guide for trees and shrubs (from Chater, C.S. and Holmes, F. W., 1975. Cooperative Extension Serice, University of Massachusetts)

TREE LIST

AILANTHUS, or Tree of Heaven (*Ailanthus*) — Cytospora canker, decay, leafspots, Nectria canker, shoestring rootrot, twig blight, Verticillium wilt.

AMUR CORK TREE (*Phellodendron*) — decay.

ALDER (*Alnus*) — canker, decay, leaf curl, powdery mildew, rust.

ALMOND (*Prunus*) — bacterial blight, brown rot, crown gall, decay, fireblight, powdery mildew, shoestring rootrot, Verticillium wilt.

ANDROMEDA (*Pieris*) — decay, dieback, leafspots, rootrot.

APPLE and CRABAPPLE (*Malus*) — bacterial blight, cankers, cedar rust, crown gall, decay, fireblight, powdery mildew, scab, shoestring rootrot, Verticillium wilt.

APRICOT (*Prunus*) — crown gall, decay, fireblight.

ARBORVITAE (*Thuja*) — blight, decay, fall browning, juniper blight, leaf blight, Pestalotia blight, shoestring rootrot, tip blight.

ASH (*Fraxinus*) — anthracnose, ash dieback, ash rust, Cytospora canker, decay, leaf scorch, leafspots, mosaic, ringspot, rust, soil (salt) pollution, Verticillium wilt.

ASH RUST is casued by the fungus: *Puccina sparganoides* (= *P. peridermiospora*). Fungus spores from ash trees cannot re-infect ash, but infect only the alternate host plants, marsh and cord grasses (*Spartina* spp.). Therefore, the disease is important near the seacoast but minor or rare inland. Spores from marsh grasses, at maturity, can infect ash trees next spring (about May 15 to June 20) when they are just coming into leaf. Especially severe in wet, rainy seasons. When the disease is visible on ash, it is too late to spray. Spraying in late May and early June with Bordeaux mixture, dichlone or ziram has given partial control. Captan, ferbam and wettable sulfur also control some rust diseases. If conditions permit, destroy nearby grass hosts. Ash cannot be well protected very near marshes containing the grasses in large quantities, and other tree species should be chosen for such locations (See also: Rusts).

ASH, MOUNTAIN or Mountain Ash (*Sorbus*) — cankers, cedar rust, crown gall, Cytospora canker, fireblight, leafspot, rusts, scab, shoestring rootrot.

ASPEN, or Poplar (*Populus*) — Cytospora canker, cankers, decays, dieback, gall, leaf blister, leafspots, powdery mildew, rusts, scab, shoestring rootrot, wetwood.

AVOCADO (*Persea*) — bacterial blight, decay.

AZALEA, or Rhododendron (*Rhododendron*) — blight, Botrytis blight, cankers, crown gall, decay, dieback, fall browning, leafspots, petal blight, powdery mildew, rootrots, rusts, shoestring rootrot, shoot blight, twig blight, walnut injury (see Pine). (The rhododendron cultivar 'English Roseum' was resistant to all 8 species of *Phytophthera* causing rootrot.)

BALD CYPRESS (*Taxodium*) — decay, twig blight.

BALSAM FIR, or Fir (*Abies*) — cankers, Cytospora canker, decay, fall browning, Juniper blight, needle blight, needlecast, rusts, shoestring rootrot, twig blight.

BARBERRY (*Berberis*) — anthracnose, bacterial leafspot, decay, rust, virus: mosaic, Verticillium wilt.

BASSWOOD, or Linden (*Tilia*) — anthracnose, canker, decay, leaf scorch, leafspots, Nectria canker, powdery mildew, sooty molds, Verticillium wilt.

BEECH (*Fagus*) — beech dark disease, bleeding canker, cankers, decay, leaf scorch or mottle, leafspots, Nectria canker, powdery mildew, wetwood, Verticillium wilt.

BEECH DARK DISEASE is caused by a combination of scale insects: *Cryptococcus fagi* and other species, followed by fungi: *Nectria coccinea faginata* and other species. It has now moved beyond New England, as far as northeastern Pennsylvania. CONTROL

TABLE 68a (continued)

of the beech scale will prevent the fungus infection. Past recommendations include dormant, miscible oil sprays, but use of oil is hazardous to beech trees. In the growing season, the white, waxy exudate protects the adults from sprays, but crawlers can be killed with dimethoate or Meta-Systox-R or malathion in August and September (see Insect Section; see also, Canker).

BIRCH (*Betula*) — canker, crown gall, decay, dieback, leaf blister, leaf rusts, Nectria canker, powdery mildew, shoestring rootrot.

BITTERSWEET (*Celastrus*) — decay.

BLACK GUM, or Tupelo (*Nyssa*) — cankers, decay, leafspots, rust, Verticillium wilt.

BLACK LOCUST, or Locust, Black (*Robinia*) — cankers, decay, leafspots, Nectria canker, powdery mildews, shoestring rootrot, virosis, Verticillium wilt, witches broom.

BOSTON IVY and VIRGINIA CREEPER (*Parthenocissus*) — canker, decay, downy mildew, leafspots, powdery mildew, rootrot, spray injury.

BOXELDER (*Acer negundo*) — included with Maple.

BOX or BOXWOOD (*Buxus*) — cankers, decay, leafspots, Nectria canker, nematodes, Phytophthora rootrot.

BUCKEYE or Horsechestnut (*Aesculus*) — anthracnose, bleeding canker, decay, leaf blotch, leaf scorch, leafspot, Nectria canker, powdery mildew, Verticillium wilt.

BUCKTHORN (*Rhamnus*) — decay, leafspots, powdery mildew, rootrot, rust, virus (mosaic).

BUTTERNUT (*Juglans*) — (see also Walnut), anthracnose, canker, decay, dieback, leafspots.

CATALPA (*Catalpa*) — anthracnose, decay, leafspots, powdery mildew, shoestring rootrot, sooty mold, Verticillium wilt.

CEDAR (Cedrus) — decay, pine twig blight, shoestring rootrot.

CEDAR, RED or Juniper (*Juniperus*) — annosus rootrot, cedar rust, decay, fall browning, juniper twig blight.

CEDAR, WHITE or White Cedar (*Chamaecyparis*) — decay, fall browning, juniper blight, rusts, witches broom.

CHERRY (*Prunus*) — bacterial blight, black knot, crown gall, decay, fireblight, gummosis, leafspot, powdery mildew, rust, shothole, twig blight, viroses, Verticillium wilt.

BLACK KNOT is caused by the fungus *Dibotryon morbosum*. Also attacks plums; may kill branches and even entire trees. CONTROL: Cut out and burn diseased branches, in fall or early spring; cut 4 to 6 inches back of the knot to remove all the fungus. This must be done for 2 or 3 years. Eradicate or treat likewise nearby diseased wild plums and cherries. Spray with wettable sulfur (3 tbsp/gal water) or captan (2 tbsp/gal water) after bloom — petal-fall for cherries; when shucks split for plums. Zineb or ferbam have also been suggested.

GUMMOSIS is not a separate disease, but a symptom caused by wounds and canker infections of cherries, peach, and plum. This exudation of a pliable gum from trees of the genus *Prunus* is equivalent to exudations of resin from conifers and of clear, watery sap from many hardwoods under like circumstances.

CHESTNUT (*Castanea*) — chestnut "blight", Cytospora canker, decay, Endothia canker, leafspot, shoestring rootrot, shothole leafspot, twig canker, Verticillium wilt.

CHESTNUT "BLIGHT" is a trunk canker caused by the fungus *Endothia parasitica*. Since the roots of chestnuts survive and send up sprouts, which in turn feed the roots, there is a continual supply of food to the tree even after its trunk has been killed by the canker. Thus, now, with relatively few of the fungal spores in the air, chestnuts often reach nut-bearing age before they are again killed back by disease. This has led to the establishment in the woods of new chestnut trees, in addition to the old sprouts. CONTROL: see Canker (it is *not* a "blight"). The infected area can sometimes be

TABLE 68a (continued)

pruned off or excised. Chinese chestnut is resistant and there exist resistant hybrids between Chinese and American chestnuts.

CHINESE SCHOLAR TREE or Japanese Pagoda Tree (*Sophora*) — canker, decay, Nectria canker, powdery mildew, rootrot, twig blight, Verticillium wilt.

CHOKECHERRY or Cherry (*Prunus*) — bacterial blight, black knot, crown gall, decay, fireblight, gummosis, leafspot, powdery mildew, rust, shothole, twig blight, viroses, Verticillium wilt.

BLACK KNOT is caused by the fungus *Dibotryon morbosum*. Also attacks plums; may kill branches and even entire trees. CONTROL: Cut out and burn diseased branches, in fall or early spring; cut 4 to 6 inches back of the knot to remove all the fungus. This must be done for 2 or 3 years. Eradicate or treat likewise nearby diseased wild plums and cherries. Spray with wettable sulfur (3 tbsp/gal water) or captan (2 tbsp/gal water) after bloom — petal-fall for cherries; when shucks split for plums. Zineb or ferbam have also been suggested.

GUMMOSIS is not a separate disease, but a symptom caused by wounds and canker infections of cherries, peach, and plum. This exudation of a pliable gum from trees of the genus *Prunus* is equivalent to exudations of resin from conifers and of clear, watery sap from many hardwoods under like circumstances.

CORK TREE (*Phellodendron*) — decay.

COTONEASTER (*Cotoneaster*) — canker, decay, fireblight, leafspots. (Do not use strepto-mycin on *C. racemiflora*, or above 65°F.)

COTTONWOOD or Poplar (*Populus*) — Cytospora canker, cankers, decays, dieback, gall, leaf blister, leafspots, powdery mildew, rusts, scab, shoestring rootrot, wetwood.

CRABAPPLE or Apple (*Malus*) — bacterial blight, cankers, cedar rust, crown gall, decay, fireblight, powdery mildew, scab, shoestring rootrot, Verticillium wilt.

CUCUMBER TREE (*Magnolia*) — decay, powdery mildew.

CURRANT (*Ribes*) — blight, blister rust of white pine, decay, leafspots.

CYPRESS (*Cypressus*) — decay.

CYPRESS, SAWARA or White Cedar (*Chamaecyparis*) — decay, fall browning, juniper blight, rusts, witches broom.

DAWN REDWOOD (*Metasequoia*) — decay.

DEUTZIA (*Deutzia*) — decay.

DOGWOOD (*Cornus*) — bleeding canker, Botrytis blight, cankers, crown canker, Cyto-spora canker, decay, leaf blight, leaf scorch, leafspots, Nectria canker, petal blight, powdery mildew, ringspot, shoestring rootrot, sooty molds, twig blights, viruses, Verticillium wilt.

DOUGLAS FIR (*Pseudotsuga*) — Botrytis blight, cankers, decay, fall browning, Juniper blight, needle blight, needle cast, rust.

ELDER (*Sambucus*) — cankers, decay, leafspots, powdery mildew, rootrots, Verticillium wilt.

ELM (*Ulmus*) — anthracnose, black leafspot, bleeding canker, cankers, Cephalosporium wilt, Cytospora canker, decay, Dothiorella wilt, Dutch elm disease, elm "scorch", leaf scorch, leafspots, mosaic, Nectria canker, phloem necrosis, powdery mildews, shoe-string rootrot, slime flux (= wetwood), sooty molds, wetwood, Verticillium wilt.

DUTCH ELM DISEASE is caused by the fungus *Ceratocystis ulmi*. Most severe on American elm. Certain elm species (such as Chinese elm and Siberian elm) and certain European selections (such as Buisman Elm and Groeneveld Elm) are resistant to all ordinary strains of the fungus. However, a more severe strain, recently discovered in England, is now known to be in many parts of Europe as well as America.

For determination of whether an elm actually has Dutch elm disease, samples from trees suspected of Dutch elm disease, taken from 3 recently wilted branches that show

TABLE 68a (continued)

dark streaking or spotting in their recent sapwood, at least 6 inches long and about 1/2 inch in diameter, labeled with name and address of collector and exact location of tree, should be mailed promptly to Shade Tree Laboratories, University of Mass., Amherst, Mass. 01002. CONTROL: (1) The best control is community-wide prevention, by suppressing the population of the elm bark beetles, that carry the disease from tree to tree, (a) by thorough destruction of the only sites where the beetles can develop from eggs to adults (under the tight, rough bark of dying or dead elm trees — or parts of elms, such as branches partly broken off in ice storms) and (b) by thorough, dormant-season spraying with methoxychlor (see Insect Section — Elm Bark Beetle). Spray must be applied early in spring, before the leaves are in the way. By *thorough* use of *both* methods in an alert and cooperative community, losses can be held as low as 1 percent of the total elm population per year. (2) This disease can also sometimes spread through roots that form natural grafts (between elms less than 50 ft. apart). Root grafts can be cut mechanically, before diseased tree is removed, or may be killed by a row of experimental sub-surface injections (18″ deep) of a soil fumigant, sodium methyl dithiocarbamate, halfway between the trees, and the holes then closed by tamping, two weeks before the infected tree is removed. (3) For many years numerous other treatments have been proposed and some even sold for prevention or therapy of this disease, including many to be injected into the tree trunks. . . .with a long history of disappointments. No cure for an injected elm and no preventive fungicide have yet been thoroughly established as satisfactory. Currently there is very active research towards these ends. Recently hope has arisen from reports of some success in retarding spread of the fungus in diseased elms by spraying foliage or injecting roots or trunks or rhizosphere with a fungicide "benomyl" but the material is still experimental and no application formulations, rates, or procedures can yet be recommended with assurance. (4) The disease can sometimes be removed from a large tree by very prompt, drastic surgery: pruning-out of the infected branch; but success is much less likely if more than 5% of the crown is wilted. Such a tree is, of course, then liable to new infections brought by more elm bark beetles from the same source. (5) Where elms are to be grown, plant resistant species or varieties except where the vase form of American is deemed essential. Resistant selections, including Buisman Elm and Groeneveld Elm, are available now from several nurseries.

Other minor wilts with identical sapwood symptoms are: Cephalosporium wilt of elm, Verticillium wilt and dieback of elm, and Acrostalagmus wilt of elm. Usually no treatment is proposed for other minor wilts, and trees often recover spontaneously only to show more or less wilting in later seasons, or later catch Dutch elm disease. (See also, general disease No. 20, Verticillium wilt.).

PHLOEM NECROSIS is caused by a mycoplasma in elm (formerly thought to be caused by the virus *Morsus ulmi*) carried by the white-banded elm leaf hopper, *Scaphoideus luteolus*. THIS DISEASE IS NOT YET CONFIRMED IN MASSACHUSETTS, but was recently discovered in the western two-thirds of New York state and northern New Jersey and repeatedly rumored in New England. A fatal disease (first kills feeding rootlets), this can also contribute to epidemics of Dutch elm disease by providing additional dead elms in which the elm bark beetles breed. Symptoms are darkened inner bark with wintergreen odor; sometimes entire foliage is small and/or yellowish as tree starves when rootlets die. CONTROL: Sanitation (prompt destruction of dead elms) becomes very important. The leaf hopper is controlled by application of two foliar sprays of methoxychlor emulsifiable concentrate as the two flushes of foliage first reach full size, the first during June, the second after mid-July. (See Insect Section — White-banded Elm Leaf Hopper.) Since elms often have the disease a year before they show symptoms, the effectiveness of spraying one year cannot be known until the fol-

286

TABLE 68a (continued)

lowing year. A tree that has already become infected, but is still symptomless, will not be benefited by the spray, although the spray may help prevent leafhoppers from feeding first on that tree and then on a healthy one. Infection of tetracycline antibiotics has been shown to repress symptoms in this mycoplasmal disease but is is not certain whether symptoms will recur.

WETWOOD is caused by the bacterium *Erwinia nimipressuralis* in elm, and by other bacterial species in poplar, maple, beech and other trees. To stop the exudation of sap (which ferments, forming a "slime flux", as it runs down the outside of the trunk) taps are installed lower on the trunk, extending far enough outwards to let the exudate drip onto the ground. A recent development is substitution of plastic tubing for metal pipes, for safety to passers-by. When the original wound ceases to flux, it can be treated like other wounds (see Wounds) and may tend to heal over. There is no known cure for wetwood in a tree, but the disease is not supposed to weaken the trunk structurally.

EMPRESS TREE (*Paulownia*) — canker, decay, leafspots, powdery mildew.

ENGLISH IVY (*Hedera*) — bacterial blight, decay, leafspots, powdery mildew, rootrot, sooty mold.

EUONYMUS (*Euonymus*) — anthracnose, crown gall, decay, leafspots, powdery mildews.

FIG (*Ficus*) — decay, crown gall.

FILBERT or Hazel (*Corylus*) — bacterial blight, black knot, crown gall, decay, leaf curl, leafspots, powdery mildew.

FIR (*Abies*) — cankers, Cytospora canker, decay, fall browning, Juniper blight, needle blight, needlecast, rusts, shoestring rootrot, twig blight,

FIR, DOUGLAS or Douglas Fir (*Pseudotsuga*) — Botrytis blight, cankers, decay, fall browning, Juniper blight, needle blight, needle cast, rust.

FIRETHORN (*Pyracantha*) — canker, decay, fireblight, rootrots, scab, shoestring rootrot, twig blight.

FLOWERING CRAB or Apple (*Malus*) — bacterial blight, cankers, cedar rust, crown gall, decay, fireblight, powdery mildew, scab, shoestring rootrot, Verticillium wilt.

FLOWERING DOGWOOD or Dogwood (*Cornus florida*) — bleeding canker, Botrytis blight, cankers, crown canker, Cytospora canker, decay, leaf blight, leaf scorch, leafspots, Nectria canker, petal blight, powdery mildew, ringspot, shoestring rootrot, sooty molds, twig blights, viruses, Verticillium wilt.

FLOWERING QUINCE (*Chaenomeles*) — brown rot and blight, canker, cedar rust, crown gall, decay, fireblight, leafspots, Nectria canker, nematodes, twig blight.

FORSYTHIA (*Forsythia*) — blight and dieback, crown gall, leafspots.

FRANKLIN TREE (*Franklinia*) — decay, leafspot.

FRINGE TREE (*Chionanthus*) — cankers, decay, leafspots, powdery mildew, galls.

GINKGO (*Ginkgo*) — decay, leafspots, (very few diseases).

GOLDENCHAIN TREE (*Laburnum*) — decay, leafspots, twig blight.

GOLDENRAIN TREE (*Koelreuteria*) — decay, leafspot, Nectria canker, rootrot, Verticillium wilt.

GRAPE (*Vitis*) — crown gall, decay, downy mildew, powdery mildew, Verticillium wilt.
DOWNY MILDEW is caused by the fungus *Plasmopara viticola* in grape. CONTROL: Apply Bordeaux mixture when mildew is discovered, and as leaves enlarge next season. Downy mildews are uncommon with trees!

GUM, BLACK or Sour Gum or Tupelo (*Nyssa*) — cankers, decay, leafspots, rust, Verticillium wilt.

HACKBERRY (*Celtis*) — decay, Ganoderma rootrot, leafspots, witches broom of hackberry.
WITCHES BROOM of hackberry is said to be caused by combination of powdery mildew and eriophyid mite. CONTROL: Sulfur, dinocap or benomyl.

TABLE 68a (continued)

HAWTHORN (*Crataegus*) — Botrytis blight, cedar rust, decay, fireblight (do not use streptomycin on *Crataegus mollis* or above 65°F), hawthorn leafspot, leafspots, powdery mildews, scab.
HAWTHORN LEAFSPOT (or "blight") is caused by the fungus *Entomosporium theumenii* (= *Diplocarpon maculatum*). It afflicts Paul's Scarlet Hawthorn and English Hawthorn, but the Washington Hawthorn (*C. phaenopyrum*) seems resistant. This is *not* a true blight, but a leafspot. Many infections on individual leaves, after wet springs, cause defoliation. (See also Leaf Spot). CONTROL: Rake up and burn old leaves in the fall, to reduce inoculum for new infections in spring and also spray new leaves several times in spring and early summer. Useful: (a) zineb at 1 1/2 lb to 100 gal water (1 1/2 tablespoonfuls/gallon) 3 or 4 times at 2-week intervals beginning the last week of April, or 3 sprays beginning in mid-June. May cause some leaf injury, or (b) cycloheximide ("Acti-dione") at 5 parts per million (read the label: if Acti-dione TGF, label says 1 oz. per 30 gal water), applied once in dry years, about June 20, or 3 times in wet years, at 10-day intervals beginning about mid-June or 3 times at 14-day intervals beginning about July 15. Cycloheximide is usually not available in most hardware and garden stores, but may be bought from handlers of golf course or greenhouse florists supplies or large farmers' outlets. See Spray injury.
HAZEL or HAZELNUT (*Corylus*) — bacterial blight, black knot, crown gall, decay, leaf curl, leafspots, powdery mildew.
HAZEL, WITCH (*Hamamelis*) — decay, leafspots, powdery mildew.
HEMLOCK (*Tsuga*) — annosus rootrot, cankers, Cytospora canker, decays, fall browning, Juniper blight, leaf blight, leaf scorch, rusts, shoestring rootrot.
HIBISCUS (*Hibiscus*) — decay.
HICKORY and PECAN (*Carya*) — anthracnose, cankers, crown gall, decay, leafspots, Nectria canker, powdery mildew, witches broom.
HOLLY and WINTERBERRY (*Ilex*) — bacterial blight, Botrytis blight, cankers, decay, infectious leafspots, leafspots from nutrient deficiency, powdery mildew, tar spot (a leafspot).
HONEYLOCUST (*Gleditsia*) — cankers, decays, leafspots, powdery mildew, rust, witches broom.
HONEYSUCKLE (*Lonicera*) — blight, crown gall, decay, leaf blight, leafspots, powdery mildews, Verticillium wilt.
HOPHORNBEAM (*Ostrya*) — cankers, decay, leaf blister, leafspots, powdery mildew, rust, shoestring rootrot.
HORNBEAM (*Carpinus*) — cankers, decay, leafspots.
HORSECHESTNUT (*Aesculus*) — anthracnose, bleeding canker, decay, leaf blotch, leaf scorch, leafspot, Nectria canker, powdery mildew, Verticillium wilt.
HYDRANGEA (*Hydrangea*) — Botrytis blight.
INKBERRY or Holly (*Ilex*) — bacterial blight, Botrytis blight, cankers, decay, leafspots, powdery mildew, tar spot.
IVY, BOSTON or Boston Ivy (*Parthenocissus*) — canker, downy mildew, decay, leafspots, powdery mildew, rootrot, spray injury.
IVY, ENGLISH or English Ivy (*Hedera*) — bacterial blight, decay, leafspots, powdery mildew, rootrot, sooty mold.
IVY, FIVE-LEAVED or Boston Ivy (*Parthenocissus*) — canker, downy mildew, decay, leafspots, powdery mildew, rootrot, spray injury.
IVY, POISON or Poison Ivy (*Rhus toxicodendron*) — decay, leafspots.
JAPANESE PAGODA TREE (*Sophora*) — canker, decay, Nectria canker, powdery mildew, rootrot, twig blight, Verticillium wilt.
JAPANESE SNOWBALL or Viburnum (*Viburnum*) — bacterial leafspot, Botrytis blight,

288

TABLE 68a (continued)

---

crown gall, decay, leafspots, powdery mildew, rusts, spray injury, Verticillium wilt. (Do *not* use sulfur on Viburnum.)

JUNIPER ("CEDAR") (*Juniperus*) — annosus rootrot, cedar rust, decay, fall browning, juniper twig blight.

KATSURA TREE (*Cercidiphyllum*) — decay, (very few diseases).

KENTUCKY COFFEE TREE (*Gymnocladus*) — decay, leafspots.

LARCH (*Larix*) — cankers, decays, juniper blight, rust, shoestring rootrot.

LAUREL or Mountain Laurel (*Kalmia*) — blight, decay, leafspots, Phomopsis blight.

LEMON (*Citrus*) — decay.

LILAC (*Syringa*) — anthracnose, bacterial blight, canker, crown gall, decay, leaf blights, leaf spots, Phytophthora blight, powdery mildew, Verticillium wilt, virus ring spot, witches broom.

LINDEN (*Tilia*) — anthracnose, canker, decay, leaf scorch, leafspots, Nectria canker, powdery mildew, sooty molds, Verticillium wilt.

LOCUST, BLACK (*Robinia*) — cankers, decay, leafspots, Nectria canker, powdery mildews, shoestring rootrot, virosis, Verticillium wilt, witches broom.

LOCUST, HONEY or Honeylocust (*Gleditsia*) — cankers, decay, leafspot, powdery mildew, rust, witches broom.

LOCUST, MORAINE or Honeylocust (*Gleditsia*) — cankers, decay, leafspot, powdery mildew, rust, witches broom.

LONDON PLANE TREE or Plane Tree (*Plantanus*) — anthracnose (more resistant than sycamore), cankers, canker-stain, decay, leafspots, powdery mildew, sooty molds, Verticillium wilt.

CANKER-STAIN OF PLANE TREE is caused by the fungus *Ceratocystis fimbriata* f. sp. *platani*. (See Canker.) NOT YET FOUND IN MASSACHUSETTS. This fungus is particularly carried by tools of humans, and appears unable to move between trees without such aid, so the most important step in its control is either to avoid all wounding of trees or to disinfest all wounds and all tools between all cuts on plane trees. The special wound dressing for this disease formerly contained a mercury fungicide. Now that mercuries are not used, benomyl is effective in tree wound paints and does not harm callus growth.

MAGNOLIA (*Magnolia*) — decay, dieback, leaf blights, leafspots, Nectria canker, powdery mildew, sooty molds, Verticillium wilt.

MAIDENHAIR TREE or Ginkgo (*Ginkgo*) — decay, leafspots, (very few diseases).

MAPLE (*Acer*) (includes Boxelder and Sycamore maple) — anthracnose, bacterial leafspot, basal canker, bleeding canker, cankers, Cytospora canker, decays, decline, Ganoderma rootrot, leaf blister, leaf scorch, leafspots, Nectria canker, nematodes, pollution of soil (salt), powdery mildew, sapstreak, scorch, shoestring rootrot, sooty molds, tar spot, viruses, wetwood (see elm), Verticillium wilt.

SAPSTREAK of sugar maple is caused by the fungus *Ceratocystis fimbriata* f. sp. *platani*. THIS DISEASE IS NOT YET FOUND IN MASSACHUSETTS, but may be so at any time because it is known in as widely separated locations as North Carolina, Michigan, and Vermont. A fatal disease that has, at least on one occasion, become locally epidemic.

MIMOSA (*Albizzia*) — decay, wilt.

MOCK ORANGE (*Philadelphus*) — decay, leafspots, Nectria canker, nematodes, powdery mildews, rust.

MORAINE LOCUST or Honeylocust (*Gleditsia*) — cankers, decays, leafspot, powdery mildew, rust, witches broom.

MOUNTAIN ASH (*Sorbus*) — cankers, cedar rust, crown gall, Cytospora canker, fireblight, leafspot, rusts, scab, shoestring rootrot.

TABLE 68a (continued)

MOUNTAIN LAUREL (*Kalmia*) — blight, decay, leafspots, Phomopsis blight.

MULBERRY (*Morus*) — bacterial blight, cankers, decay, leafspots, powdery mildew.

OAK (*Quercus*) — anthracnose, blights, cankers, chlorosis of pin oak, decays, leaf blister, leaf scorch, leaf spots, oak wilt, powdery mildews, rusts, shoestring rootrot, sooty molds, Strumella canker, twig blights, wilt.

CHLOROSIS of pin oak is caused by iron deficiency, often in alkaline soils. Iron chelates may be added to soil, special materials are needed for neutral or alkaline clay soils, but most Massachusetts soils are acidic.

OAK WILT is caused by the fungus *Ceratosystis fagacearum*. Fatal to oaks in the red and black group, but often tolerated by those in the white oak group. NOT YET FOUND IN MASSACHUSETTS, but occurs as near as central Pennsylvania. A wilt disease that in some ways plays the role among oaks that Dutch elm disease plays among elms.

OLIVE (*Olea*) — decay.

OLIVE, RUSSIAN or Russian Olive (*Elaeagnus*) — cankers, crown gall, decay, leafspots, Verticillium wilt.

ORANGE (*Citrus*) — decay, nematoses, viroses.

ORANGE, MOCK or Mock Orange (*Philadelphus*) — decay, leafspots, Nectria canker, nematodes, powdery mildew, rust.

OSAGE-ORANGE (*Maclura*) — decay, leaf blight, leafspots, rust.

PACHYSANDRA (*Pachysandra*) — blight, decay, leafspots, Volutella blight.

PAGODA TREE, JAPANESE or Japanse Pagoda Tree (*Sophora*) — canker, decay, Nectria canker, powdery mildew, rootrot, twig blight, Verticillium wilt.

PAULOWNIA (*Paulownia*) — decay.

PEACH (*Prunus*) — bacterial blight, brown rot and blight, crown gall, decay, gummosis, leaf curl, Verticillium wilt.

PEAR (*Pyrus*) — bacterial blight, cedar rust, decay, fireblight, leafspot, scab, viroses, wilt.

PECAN (*Carya*) — included with Hickory.

PERSIMMON (*Diospyros*) — canker, fireblight, Verticillium wilt.

PINE (*Pinus*) — air pollution, annosus rootrot, blister rust, cankers, decays, fall browning, juniper blight, little leaf, needle blight, needle cast, pine twig blight, red pine needle rust, rootrots, rusts, shoestring rootrot, pollution of soil (salt), sooty molds, tip blight and die-back, walnut injury, white pine blister rust.

BLISTER RUST is caused by the fungus *Cronartium ribicola;* it attacks only 5-needled pines, such as eastern white pine, where it forms cankers on branches and trunks, and may girdle them. Fungus spores produced on the pines cannot infect pines, but only gooseberry or currant bushes (*Ribes* species). From the Ribes, it cannot spread to pines more than 1 mile away, and often not more than 900 feet. CONTROL: An area-wide project to destroy nearby currant and gooseberry bushes, and to make sure that no new ones become established in the area. On pines, infected branches can be pruned off and sometimes a tree can even be saved by bark-tracing around a trunk canker, scraping off the infected surface tissues and painting the canker with a disinfectant tree-wound paint. (See also: Rusts, and Canker.)

LITTLE LEAF is caused by the soil fungus *Phytophthora cinnamomi*, which attacks the roots. NOT YET FOUND IN MASSACHUSETTS. (See Declines and Diebacks, also Decay.)

NEEDLE RUSTS OF PINE AND SPRUCE are caused by any of some 20 species in fungal genus *Coleosporium*, on red pine, pitch pine, Virginia pine, and other 2- and 3-needled pines, and by *Chrysomyxa* sp. on spruce needles. CONTROL: Destruction of wild asters and goldenrods growing near valuable pines, and spraying or dusting young pines with sulfur early in the season. Zineb or ferbam sprays have also sometimes been used to control rust diseases. (See Rusts.)

TABLE 68a (continued)

PINE-PINE GALL RUST (also known as "Western gall rust" and "Woodgate rust") is caused by the fungus *Peridermium harknessii*. It is unusual in that it needs *no* alternate host, so control must be attempted by chemical sprays, such as ferbam. This is a problem in some Christmas tree stands, where numerous galls from multiple infections make trees unsaleable.

WALNUT INJURY is caused by chemical exudate ("juglandin") from roots of walnuts and butternuts; harmless to some plants but kills others, including rhododendron and pine. CONTROL: No treatment. Grow walnut or butternuts farther away from pines, rhododendrons, etc.

PLANE TREE or London Plane (*Platanus*) — anthracnose (more resistant than sycamore), cankers, canker-stain, decay, leafspot, powdery mildew, shoestring rootrot.

CANKER-STAIN OF PLANE TREE is caused by the fungus *Ceratocystis fimbriata* f. sp. *platani*. (See Canker.) NOT YET FOUND IN MASSACHUSETTS. This fungus is particularly carried by tools of humans, and appears unable to move between trees without such aid, so the most important step in its control is either to avoid all wounding of trees or disinfest all wounds and all tools between all cuts on plane trees. The special wound dressing for this disease formerly contained a mercury fungicide. Now that mercuries are not used, benomyl is effective in tree wound paints and does not harm callus growth.

PLUM (*Prunus*) — bacterial blight, black knot (see cherry), decay, fireblight, nematosis, Verticillium wilt.

POISON IVY (*Rhus toxicodendron*) — decay, leafspot.

POPLAR (*Populus*) — Cytospora canker, cankers, decays, dieback, gall, leaf blister, leafspots, powdery mildew, rusts, scab, shoestring rootrot, wetwood.

POPLAR, TULIP or Yellow Poplar or Tuliptree (*Liriodendron*) — air pollution, cankers, decay, leaf scorch, leafspot, Nectria canker, powdery mildews, rootrot, sooty molds, Verticillium wilt.

PRIVET (*Ligustrum*) — anthracnose, canker, decay, leafspots, powdery mildew, rootrots, shoestring rootrot, twig blight, Verticillium wilt, virus variegation.

PYRACANTHA or Firethorn (*Pyracantha*) — canker, decay, fireblight, rootrots, scab, shoestring rootrot, twig blight.

QUINCE (*Cydonia*) — decay, leafspot, scab, Verticillium wilt.

QUINCE, FLOWERING (*Chaenomeles*) — brown rot and blight, canker, cedar rust, crown gall, decay, fireblight, leafspots, Nectria canker, nematodes, twig blight.

REDBUD (*Cercis*) — canker, decay, leafspots, Verticillium wilt.

RED CEDAR or Juniper (*Juniperus*) — annosus rootrot, cedar rust, decay, fall browning, juniper twig blight.

REDWOOD (*Sequoia*) — cancer, decay, Juniper blight, needle blight.

REDWOOD, DAWN (*Metasequoia*) — decay.

RETINOSPORA — decay, Juniper blight.

RHODODENDRON and AZALEA (*Rhododendron*) — blight, Botrytis blight, cankers, crown gall, decay, dieback, fall browning, leafspots, petal blight, powdery mildew, rootrots, rusts, shoestring rootrot, shoot blight, twig blight, walnut injury (see Pine). (The rhododendron cultivar 'English Roseum' was resistant to all 8 species of *Phytophthera* causing rootrot.)

ROSE (*Rosa*) — bacterial blight, bacterial leafspot, black leaf spot, Botrytis blight, cankers, crown gall, Cytospora canker, decay, fireblight, leafspots, Nectria canker, powdery mildew, nematode (root knot), rust, sooty molds, streak, Verticillium wilt, viroses (mosaic).

RUSSIAN OLIVE (*Elaeagnus*) — cankers, crown gall, decay, leafspots, Verticillium wilt.

SASSAFRAS (*Sassafras*) — cankers, curly top virus, decay, leafspots, mycoplasma: aster

TABLE 68a (continued)

yellows (see viroses; and witches broom), Nectria canker, powdery mildew, shoestring rootrot, Verticillium wilt.

SERVICEBERRY or Shadbush (*Amelanchier*) — cedar rust, decay, fireblight, leaf blight, rust, witches broom.

SHADBUSH or Serviceberry (*Amelanchier*) — cedar rust, decay, fireblight, leaf blight, rust, witches broom.

SILK TREE or Mimosa (*Albizzia*) — decay, Verticillium wilt.

SILVERBELL (*Halesia*) — decay, leafspot.

SMOKE TREE (*Cotinus*) — decay, leafspot, rusts, smothering, Verticillium wilt.

SNOWBALL, JAPANESE or Viburnum (*Viburnum*) — bacterial leafspot, Botrytis blight, crown gall, decay, leafspots, powdery mildew, rusts, spray injury, Verticillium wilt. (Do not use sulfur on Viburnum.)

SNOWBELL (*Styrax*)

SORREL TREE (*Oxydendrum*) — leafspots, decay, twig blight.

SOURGUM or Tupelo (*Nyssa*) — cankers, decay, leafspot, rust, Verticillium wilt.

SOURWOOD (*Oxydendrum*) — decay.

SPIREA (*Spirea*) — fireblight, decay, leafspot, powdery mildews, rootrot, Verticillium wilt.

SPRUCE (*Picea*) — decay, Cytospora canker, fall browning, needle casts, needle rust, rust, shoestring rootrot.

NEEDLE RUSTS OF PINE AND SPRUCE are caused by any of some 20 species in fungal genus *Coleosporium*, on red pine, pitch pine, Virginia pine, and other 2- and 3-needled pines, and by *Chrysomyxa* sp. on spruce needles. CONTROL: Destruction of wild asters and goldenrods growing near valuable trees and spraying or dusting young trees with sulfur early in the season. Zineb or ferbam sprays have also sometimes been used to control rust diseases. (See Rusts.)

STEWARTIA (*Stewartia*) — decay.

SUMAC (*Rhus*) — cankers, decay, leafspots, powdery mildew, rusts, Verticillium wilt.

SWEET GUM (*Liquidambar*) — bleeding canker, blight, dieback, decay, leafspots.

SYCAMORE and PLANE TREE (*Plantanus*) — anthracnose, canker-stain of plane tree, cankers, decay, leafspots, powdery mildew, shoestring rootrot.

SYCAMORE MAPLE (*Acer pseudoplatanus*) — included with Maple.

TAMARISK (*Tamarix*) — cankers, decay, powdery mildew, rootrot.

THORNAPPLE or Hawthorn (*Crataegus*) — Botrytis blight, cedar rust, decay, fireblight, (do not use streptomycin on *Crataegus mollis* or above 65°F), hawthorn leafspot, leafspots, powdery mildew, scab.

TREE OF HEAVEN (*Ailanthus*) — Cytospora canker, leafspots, Nectria canker, shoestring rootrot, twig blight, Verticillium wilt.

TULIPTREE (*Liriodendron*) — air pollution, cankers, decay, leaf scorch, leafspot, Nectria canker, powdery mildews, rootrot, sooty molds, Verticillium wilt.

TULIP POPLAR or Tuliptree (*Liriodendron*) — air pollution, cankers, decay, leaf scorch, leafspot, Nectria canker, powdery mildews, rootrot, sooty molds, Verticillium wilt.

TUPELO (*Nyssa*) — cankers, decay, leafspots, rust, Verticillium wilt.

UMBRELLA-PINE (*Sciadopitys*) — decay.

VIRBURNUM (*Virburnum*) — bacterial leafspot, Botrytis blight, crown gall, decay, leafspots, powdery mildews, rusts, spray injury, wilt.

VIRGINIA CREEPER or Boston Ivy (*Parthenocissus*) — canker, decay, downy mildew, leafspots, powdery mildew, rootrot, spray injury.

WALNUT and BUTTERNUT (*Juglans*) — anthracnose, bacterial blight, cankers, crown gall, fireblight, leaf blotch, leafspots, Nectria canker, Verticillium wilt, witches broom. (Can injure some other trees if too near.)

TABLE 68a (continued)

WAYFARING TREE or Viburnum (*Viburnum*) — bacterial leafspot, Botrytis blight, crown gall, decay, leafspots, powdery mildews, rusts, spray injury, Verticillium wilt.

WHITE CEDAR (*Chamaecyparis*) — decay, fall browning, juniper blight, rusts, witches broom.

WILLOW (*Salix*) — bacterial twig blight, black canker, cankers, crown gall, Cytospora canker, decay, leafspots, powdery mildew, rusts, scab ("blight"), shoestring rootrot, tar leafspot, witches broom.

WINTERBERRY or Holly (*Ilex*) — bacterial blight, Botrytis blight, cankers, decay, leafspots, powdery mildew, tar spot.

WISTERIA (*Wisteria*) — crown gall, decay, leafspots, Nectria canker, powdery mildew, rootrot.

WITCH HAZEL (*Hamamelis*) — decay, leafspots, powdery mildew.

WOODBINE or Boston Ivy (*Parthenocissus*) — canker, decay, downy mildew, leafspots, powdery mildew, rootrot, spray injury.

YELLOW POPLAR or Tulip poplar or Tuliptree (*Liriodendron*) — air pollution, cankers, decay, leafspot, Nectria canker, powdery mildews, rootrot, scorch, sooty molds, Verticillium wilt.

YELLOWWOOD (*Cladastris*) — canker, decay, powdery mildew, Verticillium wilt (only 4 diseases are known on yellowwood.)

YEW (*Taxus*) — decay, fall browning, Juniper blight, leaf scorch, leafspot, needle blight, rootrots, shoestring rootrot, twig blights.

ZELKOVA (*Zelkova*) — decay, Dutch elm disease, Nectria canker.

---

nutrition, attack of parasites and so forth. Environmental factors such as soil, weather, temperature, humidity, light and irradiation of the sun can increase or reduce the disposition of pests and also the effectiveness of poisons (Table 68a,b).

Secondary and after-effects of chemical pesticides

Should the dose of poison have no lethal effect on the pest a process of selection takes place. The resistant pests remain. The poison does not harm them any more, so its quantity or strength must be increased. Together with a particular pest, other parts of the fauna and even enemies of the pest will also be hit and destroyed. Steiner (1957) found that of 1000 arthropods living in an apple tree only about 300 are pests. 300 others are enemies of the pests, 200 are predators and 200 belong to the honey-dew seeking saprophytes, but all of them will be destroyed by the poison!

Chemical poisons can increase pest calamities by directly stimulating the multiplication of other pests. Fishes and birds will be hit directly or indirectly. The propagation of surviving birds is severely obstructed. Damage will also be done to other cultivated plants, especially through sulphuric and cupriferous compounds and herbicides, and finally human health is endangered by the use of poisons and by their residues.

It should be realized that the whole of the biocenosis is implicated in the

TABLE 68b

Control for insect and mite pests (from Chater, C.S. and Holmes, F.W., 1975. Cooperative Extension Service, University of Massachusetts)

Pest	Where to treat	When to treat						Materials for use
		Apr	May	Jun	Jul	Aug	Sep	
Aphids many kinds many colors	Crevices of bark on twigs of most deciduous plants	—						60 or 70 s oil or ethion oil
"	Undersides of leaves of elm, linden, maple, oak and many other plants; also twigs; also needles of some conifers		—	—	—			Dimethoate or Meta-Systox-R or malathion or diazinon; endosulfan permitted only where it will not drift to residential areas
Arborvitae leafminer *Argyresthia thuiella*	All foliage and soil surface			—				Carbaryl
Arborvitae weevil *Phyllobius intrusus*	Foliage and soil surface of arborvitae, chamaecyparis and juniper		—	—				Chlordane
Azalea bark scale *Erioccus azaleae*	Bark of twigs and branches of azalea, rhododendron, andromeda	—						60 or 70 s oil or ethion oil
"	"			—	—			Malathion, dimethoate, Meta-Systox-R or diazinon
Azalea leafminer *Gracilaria azaleella*	Foliage, larvae within rolled leaves of azalea			—	—	—		Diazinon or malathion or dimethoate
Balsam twig aphid *Mindarus abietinus*	Twigs of balsam fir and spruce; spray on warm days		—					Dimethoate or malathion

TABLE 68b (continued)

Pest	Where to treat	When to treat						Materials for use
		Apr	May	Jun	Jul	Aug	Sep	
Beech scale *Cryptococcus fagi*	Bark of trunk and branches					—	—	Dimethoate or Meta-Systox-R or malathion
Birch leafminer *Fenusa pusilla*	Surface of soil beneath trees to drip line		—		—			Di-Syston
Black turpentine beetle *Dendroctonus terebrans*	Entire lower few feet of trunk (special publication available upon request)		—					Lindane 1%
Boxelder bug *Leptocoris trivittatus*	Foliage and twigs			—				Carbaryl
Boxwood leafminer *Monarthropalpus buxi*	Undersides of leaves of boxwood			—				Diazinon or malathion
Boxwood psyllid *Psylla buxi*	Undersides of leaves of boxwood			—				Malathion
Bronze birch borer *Agrilus anxius*	Bark of uppermost branches especially; also bark of trunk and main stem, esp. white and European white birch			—				Lindane
Cankerworms *Alsophila pometaria, Paleacrita*	Foliage of elms, oaks, lindens, beech and		—					Carbaryl or Imidan or *Bacillus thuringiensis* or

Pest	Where found / host					Control
Clover mite *Bryobia praetiosa*	Foliage and twigs of honeysuckle, elm (esp. English elm), also other plants	—	—		—	Dicofol or chlorobenzilate or diazinon
Cooley spruce gall aphid *Adelges cooleyi*	Crevices of bark on terminal twigs and bases of buds principally on blue and Norway spruce, and douglas fir	—	Also mid-October			Malathion or carbaryl or dimethoate or Meta-Systox-R; also 60 or 70 s oil or ethion oil
Cottony maple scale *Pulvinaria innumerabilis*	Twigs of maple, esp. silver; also honey locust and other deciduous trees		—	—		Malathion or carbaryl or dimethoate or Meta-Systox-R
Dogwood borer *Thamnosphecia scitula*	Bark of trunk and branches of flowering dogwood	—	—		—	Endosulfan (not permitted to drift to residential areas) or Lindane
Dogwood sawfly *Macremphytus tarsatus*	Foliage of *Cornus* sp., esp. wild, shrubby types		—			Carbaryl
Eastern spruce gall aphid *Adelges abietis* (see also "cooley...")	Crevices of bark on ends of twigs and bases of buds	—	Also mid-October			60 or 70 s oil or ethion oil or carbaryl or Meta-Systox-R
Eastern tent caterpillar *Malacosoma americanum*	Mostly foliage of wild cherry and chokecherry, also apple, cotoneaster and lilac	—		—		Carbaryl or methoxychlor or diazinon

TABLE 68b (continued)

Pest	Where to treat	Apr	May	Jun	Jul	Aug	Sep	Materials for use
Elm bark beetles *Scolytus multistriatus Hulurgopinus rufipes*	Bark of twig crotches in tops of elm trees, also bark of trunks (see white banded elm leafhopper)	—						Methoxychlor (suitable days in March and April) (see also Disease Section, Dutch elm disease)
Elm casebearer *Coleophora limosipennella*	Upper and lower surfaces of leaves of English, Scotch and American elms		—	—				Malathion or carbaryl
Elm cockscomb gall *Colopha ulmicola*	Foliage of American and slippery elm when fully expanded		—					Malathion or diazinon
Elm flea beetle *Altica ulmi*	Foliage of American elm		—	—				Carbaryl or diazinon
Elm leaf aphid *Myzocallis ulmifolii*	Undersides of leaves on elms			—	—	—		Diazinon or dimethoate or Meta-Systox-R or malathion
Elm leaf beetle *Pyrrhalta luteola*	Upper and lower surfaces of leaves on elms			—				Carbaryl or methoxychlor
Elm leafminer *Fenusa ulmi*	Foliage of English, Scotch and Camperdown elms, also American elm		—	—				Carbaryl or dimethoate or diazinon or Meta-Systox-R

When to treat

Pest	Host	Treatment
Elm spanworm *Ennomos subsignarius*	Foliage of elm, hickory, ash, oak and other hardwoods	Imidan or carbaryl or methoxychlor
Euonymus scale *Unaspis euonymi*	Twigs and stems of euonymus, bittersweet, pachysandra	60 or 70 s oil or ethion oil
,,	,,	Diazinon or dimethoate or Meta-Systox-R
European elm scale *Gossyparia spuria*	Bark of twigs, branches, main trunk of elms, esp. in crevices and on rough areas	Malathion or dimethoate or Meta-Systox-R
European fruit lecanium scale *Lecanium corni*	Bark of twigs and small branches of elm, oak, arborvitae plus many other hosts	60 or 70 s oil or ethion oil
,,	Foliage of above	Diazinon or dimethoate
European pine shoot moth *Rhyacionia buoliana*	Small area between buds on terminals and laterals	Carbaryl or dimethoate or Meta-Systox-R
,,	Foliage and bark of twigs on Scotch, Mugho, Japanese black pine	Carbaryl
European red mite *Panonychus ulmi*	Twigs and bark of crab-apple, mountain ash, English elm	60 or 70 s oil

TABLE 68b (continued)

Pest	Where to treat	When to treat						Materials for use
		Apr	May	Jun	Jul	Aug	Sep	
Panonychus ulmi	Foliage of English elm and flowering fruits		—	—	—			Dicofol or tetradifon or chlorobenzilate
Fall webworm *Hyphantria cunea*	Foliage of many deciduous trees and shrubs. A general feeder when populations are high, but usually prefers wild cherry				—	—		Carbaryl or *Bacillus thuringiensis* or diazinon
Fletcher scale *Lecanium fletcheri*	Foliage and twigs of taxus and arborvitae		—					60 or 70 s oil
„	„				—			Diazinon or dimethoate
Forest tent caterpillar *Malacosoma disstria*	Foliage of maple, elm and many other deciduous trees		—					Carbaryl or methoxychlor
Giant hornet *Vespa crabro germana*	Bark of twigs, esp. on lilac, also rhododendron, birch				—	—		Carbaryl
Golden oak scale *Asterolecanium variolsum*	Bark of twigs and branches of white chestnut and English oak; occasionally red oak			—				60 or 70 s oil or ethion oil
„	„				—			Dimethoate or Meta-Systox-R

Species	Host / Location	Treatment
Anistoa rubicunda	Foliage of maples. Eggs may be laid over extended period (June to August)	Carbaryl
Gypsy moth *Porthetria dispar*	Foliage of oaks, elm, linden, maples, pines, and many other hardwoods	Carbaryl or Imidan or *Bacillus thuringiensis*
Hackberry psylla *Pachypsylla* spp.	Bark or twigs, branches of Celtis	Diazinon or carbaryl or malathion
Hawthorn leafminer *Lithocolletis cratagella*	Upper surface of leaves of ornamental Crataegus	Carbaryl or dimethoate or diazinon or Meta-Systox-R
Hemlock eriophyid mite *Nalepella tsugifolia*	Upper and under surfaces of needles on hemlocks	60 or 70 s oil
"	"	Meta-Systox-R
Hemlock fiorinia scale *Fiorinia externa*	Foliage of hemlocks, occasionally spruce	60 or 70 s oil plus ethion
"	"	Dimethoate or carbaryl
Hemlock looper *Lambdina fiscellaria*	Foliage of hemlock	Carbaryl or methoxychlor
Hickory leaf stem gall aphid *Phylloxera caryaecaulis*	Twigs and buds of hickory before buds develop	Diazinon or dimethoate or Meta-Systox-R
Holly leafminer *Phytomyza ilicis*	Upper and undersides of leaves on American, English holly (see native holly leaf miner)	Dimethoate or Meta-Systox-R

TABLE 68b (continued)

Pest	Where to treat	When to treat						Materials for use
		Apr	May	Jun	Jul	Aug	Sep	
Holly scale *Aspidiotus britannicus*	Leaves and twigs of hollies	—	—					60 or 70 s oil
,,	,,				—			Dimethoate or Meta-Systox-R
Honey locust mite *Eotetranychus multidigituli*	Undersides of leaves on honey locust, moraine locust				—	—		Dicofol or tetradifon
Honey locust pod gall midge *Dasyneura gleditschiae*								No effective material permitted
Imported willow leaf beetle *Plagiodera versicolora*	Upper and lower surface of leaves on most willows		—	—				Carbaryl or methoxychlor
Japanese beetle *Popillia japonica*	Foliage of very many plants including shade trees			.	—	—		Carbaryl or methoxychlor
Juniper scale *Diaspis carueli*	Foliage and twigs of junipers esp. Pfitzer, and occasionally arborvitae		—					Ethion oil or 60 or 70 s oil
,,	,,	—		—	—			Diazinon or dimethoate or Meta-Systox-R

Pest	Location / Host	Treatment
Juniper webworm *Dichomeris marginella*	Foliage of juniper, esp. upright forms	Carbaryl
Kermes oak scale *Kermes* spp.	Bark of twigs of white and red oak	60 or 70 s oil or ethion oil
"	"	Carbaryl or diazinon or Meta-Systox-R
Lacebugs *Corythuca* spp.	Undersides of leaves of walnut, ash, hawthorn, hickory, oak, linden, sycamore and similar trees	Carbaryl or Meta-Systox-R
Lacebugs *Stephanitis* spp.	Undersides of leaves on azalea, rhododendron, andromeda	Carbaryl or Meta-Systox-R
Larch casebearer *Coleophora laricella*	Foliage of European and American larch	Carbaryl
Larch sawfly *Pristiphora erichsonii*	Foliage of larches	Carbaryl
Leafhoppers (see white-banded elm leafhopper *Saphoideus luteolus*)	Undersides of leaves on most deciduous trees and shrubs and flowering fruits	Carbaryl or diazinon or Meta-Systox-R or methoxychlor
Lecanium scales *Lecanium* spp.	Twigs and small branches	60 or 70 s oil

TABLE 68b (continued)

Pest	Where to treat	When to treat						Materials for use
		Apr	May	Jun	Jul	Aug	Sep	
Lecanium scales *Lecanium* spp.	Foliage, upper and under-sides of most trees; oak, elm, maple, honey locust, flowering fuits, arborvitae, taxus				—			Diazinon or dimethoate
Lilac borer *Podosesia syringae syringae*	Bark of trunk and larger branches, esp. around wounds of lilac		—	—				Lindane or endosulfan (not permitted in residential areas)
Lilac leafminer *Gracilaria syringella*	Foliage of lilac and privet		—	—	—	—		Malathion or carbaryl
Linden looper *Erranis tiliaria*	Foliage of linden, elm, oak maple		—	—				Carbaryl
Locust borer *Magacyllene robiniae*	Bark of trunk and largest branches of old trees (one spray every 3 years)					—	—	Lindane or endosulfan (not permitted in residential areas)
Locust leafminer *Xenochalepus dorsalis*	Foliage of black locust		—	—	—			Diazinon, dimethoate or malathion
Magnolia scale *Neolecanium cornuparvum*	Bark of twigs and branches of magnolia	—		Also March				Ethion oil or 60 to 70 s oil
"	"	—				—		Carbaryl
Maple gall mites	Trunk, twigs and buds of maples, esp. silver, cut leaf	—					—	60 or 70 s oil or ethion oil

Pest	Location / Host	Control
Epinotia acericella	Undersides of leaves on sugar maple	Carbaryl, methoxychlor or diazinon
Mottled willow borer *Cryptorhynchus lapathi*	Bark of trunk and branches	Lindane or endosulfan (not permitted for use in residential areas)
Mountain ash sawfly *Pristiphora geniculata*	Foliage of mountain ash	Carbaryl or methoxychlor
Nantucket pine tip moth *Rhyacionia frustrana*	Foliage of pine, mostly pitch and black	Dimethoate or carbaryl or Meta-Systox-R
Native holly leaf miner *Phytomyza ilicicola*	Foliage of inkberry	Dimethoate or Meta-Systox-R
Oak blotch leaf miner *Cameraria* spp.	Foliage of oaks, esp. white oak, also red oak	Carbaryl or malathion
Oak leaf tier *Arbyrotoxa semipurpurana*	New opening buds of oaks as larvae hatch	Carbaryl or diazinon
Oak red mite *Oligonychus bicolor*	Uppersides of leaves of oaks; esp. northern red; also some other hardwoods	Dicofol or tetradifon
Oak skeletonizer *Buccalatrix ainsliella*	Foliage of red oak group. Two applications at 10-day intervals may be needed to control severe infestations	Carbaryl

TABLE 68b (continued)

Pest	Where to treat	When to treat						Materials for use
		Apr	May	Jun	Jul	Aug	Sep	
Oak twig pruner *Hupermallus villosus*	Infested branches				—	—	—	Pick up or prune infested branches and burn promptly
Oriental moth *Cnidocampa flavescens*	Foliage of many hardwoods, mostly Norway maple, sycamore, apple and hawthorn				—			Carbaryl
Oystershell scale *Lepidosaphes ulmi*	Bark of twigs and branches on willow, lilac, apple, ash and most trees; also pachysandra	—						60 or 70 s oil
"	"			—				Dimethoate or Meta-Systox-R
Pales weevil *Hylobius pales*	Tender bark of seedling white pine and other conifers up to 18″ high; also twigs of some larger conifers				—	—	—	Lindane
Periodical cicada *Magicida septendecim*	Bark of twigs on deciduous shade trees				—	—		Carbaryl
Pine bark aphid *Pineus strobi*	Bark of trunk, branches, twigs on white pine; also Scotch and Austrian pine	—						60 or 70 s oil
"	"	—	—					Malathion or dimethoate or

Pest	Host / Location	Treatment
Lambdina athasaria pelucidaria	Foliage of pitch pine	Carbaryl
Pine needle miner *Exoteleia pinifoliella*	Foliage of pitch pine and jack pine	None suggested
Pine needle scale *Phenacaspis pinifoliae*	Needles of white, red, Scotch, Austrian and Mugho pine	60 or 70 s oil or ethion oil
”	”	Dimethoate or Meta-Systox-R
Pine root collar weevil *Hylobius radicis*	Bark at base of trunk and soil surface 8″ out from trunk of Scotch, red, and Austrian pine	Lindane
Pine sawflies *Diprion, Neodiprion* spp.	Needles of most conifers, esp. red, Scotch, white pine and spruce	Carbaryl
Pine spittlebug *Aphorphora parallela Aphorphora saratogensis*	Foliage (needles) — need long residual — esp. white pine, also Scotch, red, Japanese red pine	Methoxychlor
Pine tortoise scale *Toumeyella numismaticum Toumeyella pine*	Twigs on Scotch, Austrian, jack and Mugho pine	Diazinon or dimethoate or Meta-Systox-R
Pine tube moth *Argyrotaenia pinatubana*	Foliage (needles) of white pine	Carbaryl
Pine webworm *Tetralopha*	Foliage (needles) on red,	Carbaryl

TABLE 68b (continued)

Pest	Where to treat	When to treat						Materials for use
		Apr	May	Jun	Jul	Aug	Sep	
robustella	pitch, white pine, usually on seedlings and in plantations							
Poplar and willow borer *Sternochaetus lapathi*	Bark of trunk and branches					—	—	Lindane, or endosulfan (not permitted near residential areas)
Red-headed pine sawfly *Neodiprion lecontei*	Thorough wetting of needles. Broods may occur throughout the season			—	—	—	—	Carbaryl
Rhododendron borer *Ramosia rhododendri*	Bark of trunk and branches		—	—				Lindane or endosulfan (not permitted near residential areas)
Rose chafer *Macrodactylus subspinosus*	Thoroughly wet foliage with a residual spray on rose, peony, grape		—	—				Carbaryl
Rose slugs *Endelomyia aethiops*	Upper and lower surface of foliage esp. on rambler rose		—	—				Carbaryl
Rust mites *Eriophyidae* spp.	Upper and lower leaf surface of elm, oak, maple,			—		—	—	Carbaryl or diazinon

Pest	Host / Location	Treatment
Prionomerus calceatus	Foliage of sassafras, magnolia	Carbaryl
Satin moth *Stilpnotia salicis*	Foliage of white poplar, willow and other salicacious plants	Carbaryl
Snowball aphid *Anuraphis viburnicola*	Foliage of Viburnum	Diazinon or dimethoate or Meta-Systox-R or endosulfan (not permitted for use in residential areas)
Southern red mite *Oligonychus ilicis*	Undersides of Ilex, esp. *I. convexa*; also other broad-leaved evergreens	60 or 70 s oil
"	"	Tetradifon or Kelthane or 60 or 70's oil or oil plus ethion (except on blue spruce)
Spruce bud scale *Physokermes piceae*	Bark of twigs	60 or 70 s oil or ethion oil
"	"	Diazinon or dimethoate or Meta-Systox-R
Spruce mite *Oligonychus unuguis*	Thorough wetting of needles and bark of arborvitae, hemlock, spruce, juniper, retinospora, pines	60 or 70 s oil or ethion oil
"	"	Dicofol or tetradifon
Sugar maple borer *Glycobius speciosus*	Thorough wetting of bark, esp. near rough bark	Lindane

TABLE 68b (continued)

Pest	Where to treat	When to treat						Materials for use
		Apr	May	Jun	Jul	Aug	Sep	
Taxus bud mite *Cecidophyopsis psilaspis*	Bark of trunk and large branches		—	—				Diazinon; endosulfan (not permitted in residential areas)
Taxus mealybug *Pseudococcus cuspidatae*	Bark of trunk and large branches	—						60 or 70 s oil or ethion oil
,,	,,		—	—				Carbaryl or malathion
Taxus pulvinaria scale *Pulvinaria floccus*	Bark of twigs and branches of taxus	—						60 or 70 s oil
,,	,,			—	—			Ethion plus summer oil or diazinon or dimethoate
Taxus weevil *Brachyrhinus sulcatus*	Foliage, bark of trunk and branches and surfaces of ground beneath where beetles walk after emerging. Two treatments may be required				—	—		Chlordane — two applications at two-week intervals will provide better clean-up
Tent caterpillars *Malacasoma* spp.	Foliage immediately following bud break, on wild cherry, maple, oaks	—	—					Methoxychlor or carbaryl or *Bacillus thuringiensis*
Two-spotted spider mite *Tetranychus urticae*	Undersides of leaves on elm, linden, rose, ornamen-		—	—				Dicofol or tetradifon

Pest	Location / Damage	Control
liriodendri		or Meta-Systox-R
Tuliptree scale *Toumeyella liriodendri*	Foliage of tulip tree	Diazinon or dimethoate or Meta-Systox-R
Tussock months *Halisidota tesselaris* *Hemerocampa leucostigma*	Foliage of many deciduous trees	Carbaryl
White banded elm leafhopper *Scaphoideus luteolus*	Underside of leaves (this is the vector of phloem necrosis of elm)	Methoxychlor
White peach scale *Pseudaulacaspis pentagona*	Bark of trunks, branches and twigs of ornamental stone fruits, privet and lilac	Dimethoate or diazinon or Meta-Systox-R
White pine aphid *Cinara strobi*	Needles of white pine	60 or 70 s oil
,,	Twigs and small branches of white pine	Dimethoate or diazinon or Meta-Systox-R
White pine weevil *Pissodes strobi*	Thorough coverage on terminals of white pine, Norway and white spruce, Scotch pine and Japanese black pine	Lindane
,,	Removal and burning of infested leaders before adults emerge reduces population	Remove and burn infested leaders
Willow flea weevil *Rhynchaenus rufipes*	Upper surface of willow foliage mined by larvae	Carbaryl

TABLE 68b (continued)

Pest	Where to treat	When to treat						Materials for use
		Apr	May	Jun	Jul	Aug	Sep	
Willow twig aphid *Lachnus* spp.	Small twigs and branches of willows				—	—	—	Diazinon or dimethoate or malathion or Meta-Systox-R
Wooly beech aphid *Prociphilus imbricator*	Undersides of leaves, also bark of beech; esp. European varieties			—	—			Diazinon or malathion or dimethoate or Meta-Systox-R or endosulfan (not permitted for residential areas)
Wooly elm aphid *Eriosoma americanum*	Bark, buds and foliage of elm; migrates in June to apple, pear, quince and hawthorn		—					Malathion
Wooly larch aphid *Chermes strobilobius*	Foliage and twigs of larch and spruces; esp. red and black		—					Malathion
Wooly maple scale *Phenacoccus acericola*	Bark of twigs of sugar and other maples as buds show green	—	—					Malathion or diazinon
Zimmerman pine moth *Dioryctria zimmermani*	Bark of twigs and branches of most pines. Adults emerge between June and August to lay eggs			—	—	—		Dimethoate

process, which therefore requires radical alterations. Impoverishment of the community may favour other species multiplying themselves into dangerous densities. The controlled use of chemical poisons makes them a speedy and effective guard against noxious organisms. However, on account of their secondary effects, the intended effect should be sought in some other way. It is not the immediate success which should be decisive but rather the long-term ecological effects.

11.2. Biological pest control

In contrast to chemical pest control its biological counterpart is built up on the rules of the ecosystem. It is therefore beneficial to the environment and will save time and money. Unfortunately up to now it has mainly concentrated on pest defence in agriculture, in fruit growing and also in forestry. The control of particular tree parasites has not yet been specialized sufficiently.

In biological pest control living beings are used to control pests such as predators, episites, sponger-parasites, pathogenic fungi, bacteria, rickettsiae, protozoa and viruses, all of which can cause lethal infections (microbiological control). Finally the members of a given species may be changed genetically by sterilization so that self-annihilation will take place (Burges and Hussey, 1971; Franz, 1964; Franz and Krieg, 1976). So the aim is not to destroy pests directly but rather to provide for their enemies. Introduction of enemies would be the simplest method but it includes the necessity of exact preliminary investigations of their way of life. If incorporation should not be possible, these useful organisms may at least be set free periodically.

The use of arthropoda

This is the incorporation of effective enemies of the pests from their countries of origin, and is used mainly against various kinds of aphids, scale and smeary and green, as well as lepidoptera. Unfortunately the pests always occupy a larger area than their specific enemies. Imports of antagonists should be well controlled. Out of 390 imported entomophagous species from all over the world into the U.S.A. only 95 were entirely integrated by 1956. Quite interesting is a comparison of expenses: in California every dollar invested in chemical pest control yielded profit of 4—5 dollars, but biological control brought 6—7 times as much profit, i.e. 30 dollars (Franz and Krieg, 1976).

11.3. Microbiological pest control

Here the group of *Bacillus thuringiensis* (Krieg, 1961) has proved quite successful (Thuricide HP, Biotrol BTB, Dipel). *Bacillus thuringiensis* has a pathogenic effect only on the larvae of butterflies (Lepidoptera). It produces a crystalline toxin which dissolves in the intestine of the larva and acts as a specific bowel poison on about 150 species of lepidoptera. In the forests of Europe, the U.S.A. and Russia, *Bacillus thuringiensis* has proved successful in the control of about 15 important species of lepidoptera (including *Dendrolimus sibiricus*, *Tortrix viridana*, *Thaumetopoea pityocampa* and *Zeiraphera diniana*). The compounds may be sprayed from an aeroplane. They are not toxic to other living beings. Saw-flies on conifers can be controlled by the "Kernpolyedervirus". The aims of insect pathology are rapid effects like the ones achieved with chemical insecticides and long-term effectiveness through regulation of the population density of the pest on a level below the damage limit. Sometimes, however, the enemies of the pests may also be disturbed by the pathogens (bees, silkworm).

11.4. Autocidal control (self-annihilation)

With the aid of manipulated individuals of the same species, a genetic control is carried out, mostly by sterilization. It is done through powerful radiation (X-rays) or by chemicals (chemosterilans) such as alkalizing substances, antimetabolites, phosphorus amides, triacides, hormones, etc., but the use of these compounds always involves the danger of damage to the environment. Autocidal control also includes translocations (fractions and alterations of the set of chromosomes).

The utilization of biological agents is called a bio-technical process. It includes attractive light, repulsive light (tinfoil strips), photoperiodic operations (flash light), acoustic repellants (shots) and ultrasonics (against mice). Moreover, amongst chemical stimuli are attractants, phagostimulants which activate egg-laying or stinging, repellants, phagodeterrents, pheromones (including sexual-attractants), endohormones (including oestrogen) and plant hormones. These methods are not expensive, they do not leave harmful residues and damage neither the ecosystem nor the human being. All bio-technical methods and substances can be precisely aimed at definite pests.

11.5. Integrated pest control

This is a matter of combining several methods (Kilgore and Doutt, 1967; Franz and Krieg, 1976). It is not an attempt to kill as many pests as possible in as short a time as possible but the regulation of pest populations according to the living and active factors of the ecosystem, both animate and inanimate. The pest must not exceed the economic injury level, that is, it must not exceed the lowest population density of a pest at which economic damage may occur. This is reached if production loss caused by the pest is twice as high as the cost of pest control.

When due consideration is given to the natural rules of the ecosystem, pest management does not become a substitute for natural proceedings but a means of correcting them.

12. EVALUATION OF TREES

Estimation of the value of a tree is not as easy as it seems to be. The forester determines it according to the price of timber. This kind of evaluation is not applicable to shade and ornamental trees. Here, not the timber value matters but the tree's ecological importance, its functions for humans, its beauty and sentimental value, its historical and aesthetic importance and so forth. How does one combine these different evaluations into a single criterion that is open to inspection? (Plates 126 and 127). They differ from man to man and from situation to situation. The first difficulty arises with the evaluation of beauty and of sentimental value. These are distinctly subjective. Before they can be determined in individual cases, the functional value (sum of the objective, scientifically examinable values) should be ascertained. Kielbaso (1975b) compares the different evaluation methods so far proposed. We can

Plate 126. How can these culturally important trees be appraised? An old "step-linden tree" in Isling/Germany (Photo Pessler).

Plate 127. Oaks on a prehistoric tomb in North Germany (Photo Kuemmerlen).

agree that the value of a tree depends on a series of factors including size, species, condition and location. It is more difficult to range the various characteristics of a particular tree under one or another factor (e.g. site or condition). The individual factors must be coordinated into a system which starts from a basic value and shows increases or decreases in value on the basis of a criterion that is open to legal and other inspection.

12.1. The individual factors

Size. In the U.S.A. the measuring of trees is currently done in bhd (breast high diameter), that is 4.5 ft above the ground or basal area. However, the ASCA suggests (Chadwick, 1975) that calliper measurements follow the American Standard for nursery stock established by the AAN and approved by the American National Standard Institute (ANSI). According to these standards, measurements must be taken six inches (= 15 cm) above the ground level for trees up to and including four inches (= 10 cm) diameter, and twelve inches (= 30 cm) above ground level for larger sizes.

Kind (species). The tree species is of great importance as many characteristics depend on it, including life expectancy, resistance against pests and diseases, growth characteristics and so forth. Different countries have therefore drawn up lists of individual tree species for evaluation purposes. In the U.S.A., the International Society of Arboriculture (formerly: the International Shade Tree Conference (ISTC)) has formerly evaluated the same species differently in various regions. A tree with a basic value of 100% will only come to 80% in another region. This shows the calculation of the basic value to be the most difficult part of the evaluation. The last edition (Chadwick and Neely, 1975) eliminates these tree lists for the single regions because of the extreme difficulty of establishing relative percentage values of trees within a region. It is up to the qualified professional plant appraiser to determine the exact value or what percentage shall be taken with regard to the qualities of the tree.

Location. This means the determination of the stand of the tree in rural land, in a garden in town, in a park, in the street of a small town or of a city, or right in the centre of the city.

Site. Does the species match ecological conditions and requirements on the stand? Is the tree placed in open soil or asphalt or concrete? Are the roots overfilled with earth? Is there any root damage resulting from excavations, etc?

Importance for the environment. What do the trees do for temperature reduction, warding off air pollution, screening off noise and ugly sights, wind screening, aesthetics, etc?

Condition. The condition of the appraised tree will be investigated in terms of stem and crown surface, growth rate, maintenance, decay, dead wood, insects, diseases, age in relation to life span, obstruction by other trees and sufficient distance between each other, etc.

In the U.S.A. size, kind, condition and (since 1975) location are evaluated and the values multiplied by each other. The current value per square inch cross-sectional area at 4.5 ft above ground level is 10 dollars (1975). With respect to condition, the tree species are divided into five classes (from 100% to 20%). Chadwick (1975) emphasizes that the recommendations for values in the lower classes up to 12 inches diameter, given by the ISA (formerly the ISTC), are too small compared with the necessary costs of replacement. Therefore recommendations are made to evaluate trees up to that size with the replacement costs, and apply the formula of the ISA for trees between 13 and 40 inches in diameter. Above that an individual evaluation should take place because the values on the current ISA table have proved unrealistic for these sizes (Chadwick, 1975). Kielbaso (1975a) and also Chadwick recommend that the hitherto lacking category "location" be included in the calculation. This could be done as follows:

316

$$\text{Total value} = \text{basic value} \times \text{species value} \times \frac{\text{condition} + \text{location values}}{2}$$

or basic V × species V × condition V × location V.

It seems impossible to give a uniform basic value for all countries because the price structure is a different one in each of them. In what follows a method of evaluation will be recommended, whose basic value can easily be adjusted to the conditions prevailing in the country concerned. Proceeding from it, with the help of Tables 69 onwards, the total value of a tree can easily be found, including increases and decreases in value. Even here, however, trees up to 30 cm bhd (12 inches) (i.e. the size of still transplantable trees) should be appraised only in terms of replacement costs because the values on the tables are too small for these trees. Beginning with trees of diameter 30 cm, the calculation proceeds as follows:

Basic value. The starting point is the price for a young tree from the nursery with circumference 10—12 cm at a height of about 1 m. The cross-sectional area in this case amounts to ca. 10 cm². A cross-section of 1 cm² would then cost 1/10 of the nursery price. For example a tree with the price of twenty dollars would have a 1 cm² cross-section value of two dollars. This basic value varies with alterations in the price structure, which can easily be determined from catalogue prices.

Size of cross-section. The diameter or circumference of the trunk of the appraised tree can be converted into cross-section size with the help of Table 69. The multiplication of the basic value by the cross-section size gives the relative value of the tree.

Location/site. To the relative value calculated so far the costs of freight and planting have to be added. These are lowest in the open country and

TABLE 69

Conversion table from stem diameter and circumference to cross section

Trunk-		
Diameter (cm)	Circumference (cm)	Cross section (cm²)
30	ca 95	ca 700
35	110	960
40	125	1250
45	140	1600
50	160	1960
60	190	2650
70	225	3850
80	255	5050
90	285	6370
100	320	7900

TABLE 70

Rectification value for location

Open country	Small town	City	City centre
1.3	1.5	2	3

TABLE 71

Rectification for site

Evaluation	Accordance of species and site	Soil compaction	Root excavation
Good (none needed)	1	1	1
up to	↓	↓	↓
Very bad	0.1	0.1	0.1

TABLE 72

Rectification for condition

Value factor	Distance to other trees	Growth (crown)	Maintenance	Damage, insects, diseases
1	Correct	Good vigorous	Good	None
0.8				
0.6	↓	↓	↓	↓
0.4				
0.2				
0	Far too narrow	Without vigour	Very bad	Totally damaged

highest in city centres. They are specified by the multiplication values in Table 70. The particularities of the site produce the following values which are on a sliding scale from 1 = good to 0.1 = worthless (Table 71). The value resulting from the multiplication of relative value × location × site needs further rectification according to the condition (Table 72). A drop in value occurs every time a factor is applicable. If more than one factor is applicable, the value factor of the next lower line will be used. In this procedure the aging of the tree is allowed for. Thus the value of a tree amounts to:

Total value = size V × kind V (species) × location V × condition V.

If no value drop has occurred so far, life expectancy must now be included in the calculation. This is expressed as age-diminution value, which is calculated on the basis of a tree's past life span insofar as it extends beyond the midpoint of its normal life expectancy (compare Table 74, Age of trees). The age-diminution value is calculated according to Table 73.

318

TABLE 73

Age-diminution. Formula: At a life expectancy of, e.g., 100 years after the midpoint of total life expectancy, (culmination point), the diminution of value after ...... years amounts to ......% of the value at the culmination point (after Koch, 1971, modified) *.

	10	20	30	40	50	60	70	80	90	100 years
Linear $\dfrac{A}{L}$	10	20	30	40	50	60	70	80	90	100 %
Parabolic $\dfrac{A^2}{L^2}$	1	4	9	16	25	36	49	64	81	100 %

* A = age after the midpoint of life expectancy (life span); L = total life expectancy depending on the location, etc.

TABLE 74

Maximum age and height of some deciduous trees and conifers (according to measurements by Altmann, Beckmann, Buesgen, Dittmer, Molisch, Muench and others)

Tree species	Age (years)	Height (m)
Deciduous trees		
Acer campestre	200	10 — 15
negundo	75 — 100	20 — 25
platanoides	200 — 400	25 — 30
pseudoplatanus	600	25 — 30
rubrum	200 — 250	30 — 35
saccharinum	50 — 125	30 — 35
Aesculus glabra		25
hippocastanum	300	25 — 30
octandra	60 — 80	30
Alnus incana	50 — 100	25
glutinosa	100 — 300	25
rubra	50 — 100	25
Betula nana	80	1 — 2
papyrifera	80 — 100	35
pendula	100 — 120	25 — 30
populifera	50	15 — 20
pubescens	100 — 120	25
Carpinus betulus	150 — 250	25 — 30

TABLE 74 (continued)

Tree species	Age (years)	Height (m)
Castanea dentata	100 — 300	35
sativa	500 — 700	25 — 30
Catalpa bignonioides	100	15
speciosa	100	35
Carya glabra	200 — 300	35
cordiformis	175	25
ovata	250 — 300	35
Celtis occidentalis	75 — 150	35 — 40
Cornus florida	125	15
Corylus avellana	80	4 — 8
colurna	100	20
Diospyros virginiana	60 — 80	35 — 40
Fraxinus americana	250 — 300	35
excelsior	250	35 — 40
Fagus grandifolia	300 — 400	35
sylvatica	350	35
Gleditsia triacanthos	120	40
Hedera helix	200	
Ilex opaca	100 — 150	15 — 20
Juglans cinerea	80	15 — 20
nigra	200 — 300	30
regia	300 — 400	30
Liquidambar styraciflua	200 — 300	40
Liriodendron tulipifera	200 — 250	40 — 50
Magnolia acuminata	80 — 250	25 — 30
grandiflora	80 — 120	25
Nyssa sylvatica		25
Olea europaea	1500 — 1700	
Platanus hybrida	300 — 500	35
occidentalis	250 — 500	35 — 40
orientalis	400 — 600	30

TABLE 74 (continued)

Tree species	Age (years)	Height (m)
Populus alba	200 — 300	30
balsamifera	100 — 150	25 — 30
canescens	150	20
deltoides	60 — 100	30 — 50
nigra	300	35
sargentii	50 — 90	25 — 30
tremula	80 — 150	25
tremuloides	70 — 100	25 — 30
trichocarpa	150 — 200	35 — 40
Quercus alba	300 — 600	30
bicolor	300	20 — 25
coccinea	150	25 — 30
lyrata	300 — 400	25 — 30
macrocarpa	200 — 400	25 — 30
nigra	175	20 — 25
palustris	125 — 150	25 — 30
pedunculata	700 — 1000	40
petraea	700	40
rubra	250 — 500	25 — 35
velutina	150 — 200	10 — 15
Robinia pseudacacia	100 — 200	20 — 25
Salix alba	150	25
amygdaloides	50 — 100	12 — 18
nigra	75 — 125	15 — 30
Sassafras albidum	100 — 500	20 — 30
Sorbus aucuparia	80 — 100	15
torminalis	100 — 150	15 — 20
Tilia americana	100 — 140	25 — 40
cordata	1000	30
heterophylla	100	25 — 30
platyphyllos	700 — 1000	30
Ulmus americana	150 — 300	30 — 35
campestris	400	30
leavis	250	10 — 35
Conifers		
Abies alba	300 — 800	
amabilis	250 — 300	50 — 70
balsamea	150 — 200	18 — 25
concolor	150 — 400	35 — 55
grandis	200 — 400	45 — 75
lasiocarpa	150 — 200	30 — 40
magnifica	250 — 400	50 — 70

TABLE 74 (continued)

Tree species	Age (years)	Height (m)
Chamaecyparis lawsoniana	300 — 500	50 — 60
nootkaensis	300 — 600	25 — 35
Cedrus libani	1200 — 3000	40
Cupressus sempervirens	2000 — 6000	50
Juniperus communis	500 — 1000	
virginiana	150 — 300	15 — 30
Larix decidua	500	50
laricina	100 — 200	25 — 30
leptolepis	400 — 500	55
occidentalis	300 — 600	50 — 60
Libocedrus decurrens	300 — 400	30 — 55
Picea abies	300 — 400	30 — 50
engelmannii	200 — 500	35 — 45
glauca	150 — 350	20 — 30
mariana	150 — 250	12 — 25
pungens	150 — 350	30 — 35
sitchensis	400 — 750	60 — 80
Pinus aristata	4600	
banksiana	80 — 150	18 — 25
cembra	1200	25
contorta	120 — 300	20 — 30
echinata	200 — 300	30 — 45
flexilis	200 — 400	15 — 25
jeffreyi	300 — 500	30 — 35
lambertiana	450 — 650	60 — 65
monticola	200 — 500	35 — 40
nigra	500 — 600	30 — 50
ponderosa	200 — 500	55 — 70
resinosa	250 — 350	30 — 40
rigida	100 — 200	20 — 30
strobus	300 — 500	35 — 65
sylvestris	250 — 600	35 — 40
Pseudotsuga menziesii		60 — 80
taxifolia	500	80 — 90
Sequoiadendron giganteum	2000 — 3000	85 — 105
Sequoia sempervirens	1000 — 2500	80 — 110
Taxodium distichum	600 — 1200	35 — 45
mexicanum	2000 (Lyr)	

322

TABLE 74 (continued)

Tree species	Age (years)	Height (m)
Taxus baccata	900 — 2000	15
Thuja occidentalis	300 — 400	15 — 35
plicata	500 — 800	60 — 75
Tsuga canadensis	300 — 600	25 — 45
heterophylla	300 — 600	50 — 75
mertensiana	200 — 500	30 — 40

12.2. Calculation of partial losses

These can occur in the root area, on the trunk and on the crown. Crashing cars, breaking branches or overfillings can diminish the value. When these events have occurred the value of the whole tree prior to the acute injury has to be calculated first. Then the extent of the partial loss will be investigated and expressed as a percentage of the total value. It should be remembered that lengthwise trunk injuries do not cause as much harm as those across the trunk, so that it is necessary first to determine the breadth of the injury in relation to the total circumference of the trunk (Table 75). The breadth of destroyed cambium is always decisive.

Pulled off and broken branches. First the present value of the whole tree has to be calculated and then the extent of the damage in % of the undamaged crown. Where half of the branches are lost, this is considered total damage. Attention should be paid to conifers and other types of plant not capable of sprouting any more out of the old wood. Very old trees with a diameter of one meter and more will prove difficult to evaluate. Certainly the life expectancy of these trees nears its end, but in the case of certain trees surgical measures may extend to it to the utmost. The costs of thorough surgical

Table 75
Trunk injuries (bark and cambium)

Injury in % of trunk circumference	Diminution of value in %
Up to 20 %	At least 20
25	25
30	35
40	70
45	90
50	100

support either for the preservation of the tree, or for the removal of damage as well as for further maintenance, can be used as evaluation data.

Examples. A mighty red oak (*Quercus rubra*) with a trunk diameter of 50 cm is to be evaluated. A young tree of this species, with a circumference of 10—12 cm, costs DM 40 in Western Germany; thus 1 cm^2 costs DM 4. The size of the cross-section of the tree is 1960 cm^2. 4 × 1960 = DM 7840 gives the relative tree value. As the tree stands in an urban area amid numerous buildings its value is 1.5 times higher; DM 7840 × 1.5 = DM 11,760, but as it has been neglected and is in poor health a diminishing factor of 0.7 brings us to the final sum of DM 8232. Even though the tree is 250 years old and has already surpassed half of its life expectancy (400 years) by 50 years, no age-diminution value needs to be considered as this is already included in the evaluation of the tree's condition, but if the tree were still healthy and vigorous and well cared-for, a linear age-diminution of 25% would be entered so that the total value would come to DM 11,760 — DM 2940 = DM 8820.

If the trunk of the same tree were crashed into by a lorry, and the breadth of the bark injury was 50 cm (= 33% of the trunk circumference), the value drop would amount to 50% of the final sum of DM 8820, coming to DM 4410. For the calculation for trees with two or more trunks, the value used is based on the diameter of the main trunk plus 50—70% of the value derived from the combined diameters of the remaining trunks (Chadwick, 1975).

It is a point of discussion whether any subjective factors, or how many of them (beautiful autumnal colouring, a picturesque shape, etc.) should be included in the evaluation. However the importance of the tree for environmental protection may today be included in the evaluation of its location (Table 70). Alternatively, an altogether different evaluation may be used for this purpose as recommended by Ermer (1974). The importance of the tree for the environment depends on the dimension of its crown. In this case the costs for the volume of the crown of the appraised tree have to be determined on the basis of offers submitted for a tree of comparable crown volume of the same species. If no tree exists having an equivalent crown volume then the price and dimension of the largest possible tree in a nursery must be taken as a basis. The quotient of the total crown volume of the appraised tree and of the largest possible tree from the nursery have to be multiplied by this price:

$$\frac{CV_t}{CV_n} \times P = \text{value}$$

C = crown, V = volume, t = total, CV_t = total crown volume of the appraised tree, CV_n = crown volume of the largest possible tree from a nursery, P = price of this tree from the nursery. In addition, planting costs and rectification values, if any, must be considered in these evaluations.

REFERENCES

Adams and Ellis, 1966. Some physical and chemical changes in the soil brought about by saturation with natural gas. Proc. Soil Sci. Soc. Am., 24: 41—44.

Adamse, A.D., Hoeks, J., de Bont, J.A.M. and van Kessel, J.F., 1972. De microbiologische activiteit in de bodem rondom een aardgaslek. (6. Rapport of Studiecommissie invloed aardgas op beplantingen) 's Gravenhage, 15 pp.

Amelung, W., 1952. See Vogt, H. and Amelung, W. 1952.

Anon., 1970. Air-pollution. A ray of hope from Canada. Commonw. For. Rev., 49 (4): 302.

Anon., 1971. Weltweite Umweltverschmutzung. Umsch. Wiss. Tech., 71: 305—308.

A.R.S., 1964. Protecting trees against damage from construction work. Agric. Inf. Bull. No. 285, A.R.S. — U.S. Dept. Agric., 26 pp.

Assmann, E., 1961. Waldertragskunde. BLV München, Bonn, Wien, 490 pp.

Attoe, O.J., Rasson, F.L., Dahnke, W.C. and Boyle, J.R., 1970. Fertilizer release from pockets and its effect on tree growth. Proc. Soil Sci. Soc. Am., 34 (1): 137—141.

Banukiewicz, E., 1973. Treatment of tree cavities with Polyurethane foam. Proc. I.S.T.C., 49: 33—35 (Supp. Arborist's News 39).

Barthelmess, A., 1972. Wald — Umwelt des Menschen. Karl Alber, Freiburg, München, 332 pp.

Baumgartner, A., 1956. Untersuchungen ueber den Waerme- und Wasserhaushalt eines jungen Waldes. Ber. Dtsch. Wetterdienstes 5, No. 28.

Baumgartner, A., 1970. Sauerstoffumsätze von Baeumen und Waeldern. Allg. Forstz., 25 (22): 482—483.

Baumgartner, A., 1972. Atmosphaere und Biosphaere. Bl. Natur.- Umweltschutz, Heft, 2: 32—33.

Baumzeitung, 1971. Realisierung der im Bundesbaugesetz gegebenen Moeglichkeiten von Baumanpflanzungen und des Baumschutzes. Baumzeitung, 5: 57.

Baumzeitung, 1972. Die Baumstiftung in Holland. Baumzeitung, 6: 56—57.

Baumzeitung, 1972. Baumschutz in der Schweizerischen Gesetzgebung. Baumzeitung, 6: 58.

Baumzeitung, 1974. Aufgrabeschein (Aufgrabungen unter Bäumen). Baumzeitung, 8: 31—32.

Baumzeitung, 1975. Wiener Baumschutzgesetz. Baumzeitung, 9: 58—60.

Beardall, M.J., 1972. Tree feeding and nutrition. Arboric. Assoc. J., 2: 80—88.

Beck, G., 1967. Pflanzen als Mittel zur Laermbekaempfung. Patzer, Berlin, 92 pp.

Becker, K., 1971. Ausmass und Gefahren der Umweltverschmutzung. Baumzeitung, 5: 60—63.

Becker, K., 1974. Laerm aus der Sicht des Mediziners. Baumzeitung, 8: 63—64.

Beeching, A.G., 1969. Letters to the Editor. Arboric. Assoc. J., 1 (9): 249—250.

Beeching, A.G., 1970. Pitfalls in braching. Arboric. Assoc. J., 1 (10): 273—278.

Belot, Y. and Caput, C., 1973. Influence des ecrans vegetaux sur la dispersion et le depot au sol des polluants atmosphériques. Agriculture (Paris), no. 366: 175—177.

Berge, H., 1976. Immissionsprobleme in der Ziegelindustrie. Reinhaltung der Luft, 36: 460—465.

Bernatzky, A., 1960. Von der mittelalterlichen Stadtbefestigung zu den Wallgruenflaechen von heute. Ein Beitrag zum Gruenflaechenproblem deutscher Staedte. Patzer, Berlin, Hannover, Sarstedt, 123 pp.

Bernatzky, A., 1967. Arbeitsleistung und Wert des Baumes. Baumzeitung, 1: 3—5.

Bernatzky, A., 1968. Schutzpflanzungen zur Luftreinigung und Besserung der Umweltbedingungen. Baumzeitung, 2: 37—42.

Bernatzky, A., 1969a. Vorgaerten und Baeume. Baumzeitung, 3: 29—30.

Bernatzky, A., 1969b. Stadtklima und Baeume. Baumzeitung, 3: 43—47.

Bernatzky, A., 1969c. Die Bedeutung von Schutzpflanzungen gegen Luftverunreinigungen. Air pollution. Proc. Ist European Congress Influence of Air Pollution on Plants and Animals. PUDOC, Wageningen, 383—395.

Bernatzky, A., 1972. Gruenplanung in Baugebieten. Deutscher Fachschriftenverlag, Wiesbaden, 126 pp.

Bernatzky, A., 1973a. Sauerstoff und Baeume. Baumzeitung, 7: 5—9.

Bernatzky, A., 1973b. Baum und Mensch. Kramer, Frankfurt/M, 203 pp.

Bernatzky, A., 1973c. Erholung und Freizeit im Grünen. DBZ Nr. 7: 1359—1370.

Bernatzky, A., 1974. Trees on building sites. Landscape Planning, 1: 255—288.

Beroza, M. (Editor), 1970. Chemicals controlling insect behavior. Academic Press, New York, 170 pp.

Bichlmaier, F. and Gundermann, E., 1974. Beitraege zur Quantifizierung der Sozialfunktionen des Waldes. (Forschungsbericht No. 21 der Forstl. Forschungsanstalt München), 222 pp.

Bitzl, F., 1968. Verkehrsunfaelle im Zusammenhang mit dem Baumbestand an Strassen. In: Deutscher Rat fuer Landespflege, Heft 9: Landschaftspflege an Verkehrsstrassen, pp. 9—12.

Blum, W., 1965. Luftverunreinigung und Filterwirksamkeit des Waldes. Der Forst- und Holzwirt, 20.

Blum, W.E., 1974. Salzaufnahme durch die Wurzeln und ihre Auswirkungen. Berichte Eidgenöss. Anst. f. d. forstl. Versuchswesen No. 123: 41—44.

Boeminghaus, D., 1974. Der Baum an Landstrassen als informationspsychologische Groesse fuer den Autofahrer. Gartenamt, 23 (10): 563—574.

Bohne, H., 1964. Fluor-Emission und Tunnelofen. Staub, 24: 261—264.

Bohne, H., 1972. Klaerung eines Rauchschadensfalles bei Kiefernbestaenden im Ruhrgebiet. Mitt. Forstl. Bundes-Versuchsanst. Wien, No. 97: 141—150.

Bonnemann, A. and Röhrig, E., 1971. Der Wald als Vegetationstyp und seine Bedeutung fuer den Menschen. Vol. I. Waldbau auf oekologischer Grundlage. Parey, Hamburg, Berlin, 229 pp.

Bonner, I. and Galston, A., 1952. Principles of Plant Physiology. San Francisco.

Bovay, E., 1972. The effects of air pollution on plants, animals and soil. Nature in Focus, No. 13: 6—8.

Brahe, P., 1971. Ergebnisse einer Befragung ueber bepflanzte Parkplaetze. Gartenamt, 20: 219—230.

Brahe, P., 1974. Klimatische Auswirkungen von Gehoelzen auf umbauten Stadtplaetzen. Gartenamt, 23: 61—70.

Brahe, P., 1975. Zur Bepflanzung von Parkplaetzen in staedtischen Bereichen Oekologische Voraussetzungen und Auswirkungen sowie planerische Anwendung. Doctoral Thesis, T.H. Aachen, 131 pp. + 36 figs, 7 tables.

Brandt, C.J. and Rhoades, R.W., 1972. Effects of limestone dust accumulation on composition of a forest community. Environ. Pollut., 3: 217—225.

Brandt, C.J. and Rhoades, R.W., 1973. Effects of limestone dust accumulation on lateral growth of forest trees. Environ. Pollut., (4/3): 207—213.

Braun, G., 1974. Der Wald als Indikator und Schutzvegetation. Forstwiss. Centralbl., 93: 91—98.

Braun-Blanquet, J., 1932. Plant sociology: the study of plant communities. (Translated and edited by G.D. Fuller and H.C. Conard). McGraw-Hill, New York.

Braun-Blanquet, J., 1964. Pflanzensoziologie, 3. Aufl. Springer, Wien, 865 pp.

Bridgemann, P.H., 1976. Tree surgery. A complete guide. David and Charles, Newton Abbot, 144 pp.

Brockmann, C.F., 1959. Recreational use of wild lands. New York.

Broecker, W.S., 1970a. Man's oxygen reserves. Science, 168 (2): 1537—1538.

Broecker, W.S., 1970b. Gas exchange. Environment, 12: 42—43.

Brown, C.L., 1954. The response of shortleaf and loblolly pines to micronutrient deficiencies of zinc, boron, manganese and copper. M.S. Thesis, University of Georgia, 63 pp.

Brown, C.E., 1972. The pruning of trees, shrubs and conifers. Faber and Faber, London, 351 pp.

Bruenig, E.F., 1971. Die Sauerstofflieferung aus den Waeldern der Erde. Forstarchiv, 42 (2): 21—23.

Bucher, J.B., 1975. Zur Phytotoxizitaet der nitrosen Gase. Eine Literaturuebersicht. Schweiz. Z. Forstwes., 126 (5): 373—391. (= Ber. Eidg. Anst. Forstl. Versuchwesen, No. 140 Birmensdorf, 20 pp.).

Buchholz, E., 1950. Der Kampf gegen die Duerre in der Sowjetunion. Mitt. Bundesforschungsanst. Forst- Holzwirtsch. No. 12: 1—38.

Bukovac, M.J. and Witwer, 1957. Absorption and mobility of foliar-applied nutrients. Plant Physiol., 32 (5): 428—435.

Burges, H.D. and Hussey, N.W. (Editor) 1971. Microbial control of insects and mites. Academic Press, London, 861 pp.

Buschbom, U., 1967. Chlorideinwirkungen auf oberirdische Sprossteile von Holzgewaechsen. Doct. Thesis, University Goettingen.

Buschbohm, U., 1972. Salzschaeden an Strassengehoelzen. Staedtehygiene, 1972: 48—51.

Butin, H. and Zycha, H., 1973. Forstpathologie fuer Studium und Praxis. Thieme, Stuttgart, 117 pp.

Caborn, J.M., 1965. Shelterbelts and windbreaks. Faber and Faber, London, 288 pp.

Chadwick, L.C., 1975. ASCA recommendations for modification of the ISTC shade tree evaluation formula. J. Arboric., 1 (2): 35—38.

Chadwick, L.C. and Neely, D., 1975. A guide to the professional evaluation of landscape trees, specimen shrubs and evergreens. Int. Soc. Arboric., Urbana, Ill., 18 pp.

Chandler, T.J., 1965. The climate of London. Hutchinson, London, 292 pp.

Chater, C.S. and Holmes, F.W., 1975. Insect and diseases control guide for trees and shrubs. Cooperative Extension Service University of Massachusetts, 63 pp.

Chevron Chemical Company, without year. Trees for a more liable environment, San Francisco, 20 pp.

Chirkova, T.V., 1968. Oxygen supply to the roots of certain woody plants kept under anaerobic conditions. Sov. Plant Physiol., 15: 475—477.

Chrometzka, P., 1974a. Salztoleranz, Ursachen und praktische Moeglichkeiten zu deren Steigerung. Eur. J. For. Pathol., 4 (1): 50—52.

Chrometzka, P., 1974b. Salzvertraeglichkeit bei Pflanzen. Dtsch. Baumsch., 26: 256—257.

Chrometzka, P., Wagner, A. and Reinshagen, A., 1973. Gegen Streusalz tolerante Gehoelze. Muellkompost zur Verbesserung der Wachstumsbedingungen? Garten Landschaft, 83: 509—511.

Cole, L.C., 1970. Gas-exchange. Environment, 12: 40—42.

Cook, D.I. and Haverbeke, D.F. van, 1971. Trees and shrubs for noise abatement. Lincoln, Nebraska: Rocky Mountain Forest and Range Exp. Stn., East Campus, Univ. Nebraska, 77 pp.

Cook, D.I. and Haverbeke, D.F. van, 1972. Trees, shrubs and landforms for noise control. J. Soil Water Conserv., 27 (6): 259—261.

Coombe, D.E., 1966. The seasonal light climate and plant growth in a Cambridgeshire wood. In: R. Bainbridge, G.C. Evans and D. Rackham, Light as an Ecological Factor. Oxford.

Crawford, R.M.M., 1974. Tree-root survival under flooding, concrete, traffic and gas leakage. In: Wye College, Tree growth in the landscape, pp. 10—13.

Daessler, H.G., Ranft, H. and Rehn, K.H., 1972. Zur Widerstandsfaehigkeit von Gehoelzen gegenueber Fluorverbindungen und Schwefeldioxid. Flora, 161: 289—302.

Dale, J., McComb, A.L. and Loomis, W.E., 1955. Chlorosis, Mycorrhiza and the growth of pines on a high-lime soil. For. Sci., 1: 148—157.

Daniels, R., 1974. Salt: ice-free walks and dead plants. Arborist's News, 39: 13—15.

Darley, E.F., 1969. The role of photochemical air pollution on vegetation. Air-pollution. Proc. Ist European Congress Influence of Air Pollution on Plants and Animals. PUDOC, Wageningen, 137—142.

Davies, W.J., Kozlowski, T.T., Chaney, W.R. and Lee, K., 1972. Effects of transplanting on physiological responses and growth of shade trees. Proc. ISTC, 48: 22—30.

Davison, A.W., 1971. The effects of de-icing salt on roadside verges. I. Soil and plant analysis. J. Appl. Ecol., 8: 555—561.

Davitaya, F., 1971. Verschmutzung der Erdatmosphaere und Veraenderung ihrer Gaszusammensetzung. Izv. Akad. Nauk SSSR (Rep. Acad. Sci. Soviet Socialist Republics, Geogr. Series No. 4) German summary in: Ideen des exakten Wissens 1972: 559.

Dengler, A., 1971. Waldbau auf oekologischer Grundlage. see: Bonnemann, A. and Röhrig, E.

Deutscher Rat fuer Landespflege (ed). 1968. Landschaftspflege an Verkehrsstrassen Heft No. 9. Bonn-Bad Godesberg, 54 pp.

Dochinger, L.S., 1971. The symptoms of air pollution injuries to broadleaved forest trees. Mitt. Forstl. Bundes-Versuchsanst. Wien, No. 92: 7—32.

Dochinger, L.S., 1972. Can trees cleanse the air of particulate pollution? Proc. ISTC, 48: 45—48. (Supplement to Arborist's News, 38 (1973) No. 1.)

Dunball, A.P., 1968. The practical problems of motorway planting. Arboric. Assoc. J., 1 (7): 179—186.

Dunball, A.P., 1972. The landscape treatment of trunk roads and motorways. Arboric. Assoc. J., 2 (2): 38—42.

Dunger, W., 1972. Untersuchungen zur Langzeitwirkung von Industrie—Emissionen auf Boeden, Vegetation und Bodenfauna des Neissetales bei Ostritz/Oberlausitz. Abh. Ber. Naturkundemus. Goerlitz, 47: 1—40.

Dyring, E., 1973. The principles of remote sensing. Ambio, 2 (3): 57—69.

Eaton, F.M., Olmstead, W.R. and Taylor, O.C., 1971. Salt injury to plants with special reference to cations versus anions and ion activities. Plant Soil, 35: 533—547.

Ehwaldt, E., 1957. Ueber den Naehrstoffkreislauf des Waldes. Sitzungsber. Dtsch. Akad. Landwirtschaftswiss. Berlin, 6, 56 pp.

Eimern, J. van, Karschon, R., Razumova, L.A. and Robertson, G.W., 1964. Windbreaks and shelterbelts. WMO Techn. Note No. 59, Geneva, 188 pp.

Eliade, M., 1954. Die Religionen und das Heilige. (Französ. Originalausgabe: Traité d'histoire des Religions) O. Mueller, Salzburg, 600 pp.

Ellenberg, H., 1956. Grundlagen der Vegetationsgliederung, 1. Teil: Aufgaben und Methoden der Vegetationskunde. Ulmer, Stuttgart, 136 pp.

Ellenberg, H. (Editor), 1971. Integrated experimental ecology. Methods and results of ecosystem research in the German Solling Project. (Ecological Studies Vol. II). Springer, Berlin, Heidelberg, New York, 214 pp.

Emschermann, Dr., 1973. Zu Imissions- und Streusalzschaeden im Strassenbegleitgruen. Neue Landschaft, (1973): 12—15.

Enderlein, H. and Stein, G., 1964. Der Saeurestand der Humusauflage in den rauchgeschaedigten Kiefernbestaenden des staatlichen Forstwirtschaftsbetriebes Duebener Heide. Arch. Forstwes., 13: 1181—1191.

Ermer, K., 1974. Verfahren zur Wertberechnung von Strassenbaeumen. Gartenamt, 23 (10): 574—577.

Fenska, R.R., 1964. The complete modern tree experts manual. Dodd, Mead, New York, 345 pp.

Ferry, B.W., Baddeley, M.S. and Hawksworth, D.L., 1973. Air pollution and lichens. Athlone Press, Univ. London, 389 pp.

Fiedler, H.H.J. and Hoehne, H., 1965. Vorkommen und Gehalt der Makro-Naehrstoffe in Waldbaeumen. Wiss. Z. Tech. Univ. Dresden, 14: 989—999.

Flemming, G., 1967a. In welchem Fall können Waldstreifen die Rauchgaskonzentration vermindern? Luft-Kältetech.: 255—258.

Flemming, G., 1967b. Concerning the effect of terrain configuration on smoke dispersal. Atmospheric Environment, Vol. 1, Pergamon, 239—252.

Flemming, G., 1972a. Qualitative Modellbetrachtungen zur Staubschutzwirkung von Gehoelzen. Arch. Naturschutz. Landschaftsforsch., 12: 177—188.

Flemming, G., 1972b. Die Bedeutung des Waldklimas fuer die Erholungswirkung des Waldes. Soz. Forstwirtsch., 22: 55—57.

Fontaine, R.G., 1972. Forestry and environment. Geoforum, No. 10: 71—80.

Forstl. Bundesversuchsanstalt Wien (Ed.), 1971. Methods for the identification and evaluation of air pollutants injuries to forests. Mitt. Forstl. Bundes-Versuchsanst. Wien, 92: 271 pp.

Forstl. Bundesversuchsanstalt Wien (Ed.), 1972. Wirkungen von Luftverunreinigungen auf Waldbaeume. Mitt. Forstl. Bundes-Versuchsanst. Wien, 97: 646 pp.

Foss, B. and Nehls, S., 1973. Road salt damages trees. Wis. Conserv. Bull., 38: 22—23.

Foy, C.L., 1970. Plants and pollution. Agronomy Health, 16: 37—59. .

Franz, J.M., 1964. Forest insect control by biological measure. FAO/IUFRO Symp. Internationally Dangerous Forest Diseases and Insects. Oxford, 21 pp.

Franz, J.M. and Krieg, A., 1976. Biologische Schaedlingsbekaempfung. Parey, Hamburg, Berlin, 208 pp.

Fritts, H.C., Blasing, T.J., Hayden, B.P. and Kutzbach, J.E. 1971. Multivariate techniques for specifying tree-growth and climate relationships and for reconstruction anomalies in paleoclimate. J. Appl. Meteorol., 10: 845—864.

Fuggle, R.F. and Oke, T.R., 1970. Infrared flux divergence and the urban heat island. In: Urban climates, Techn. Note No. 108, WMO Geneva, pp. 70—77.

Garber, K., 1967. Luftverunreinigungen und ihre Wirkungen. Borntraeger, Berlin, 279 pp.

Garber, K., 1973. Luftverunreinigungen, eine Literaturuebersicht. (= Bericht No. 102 Eidgeno. Anst. Forstl. Versuchswesen) Birmensdorf, 216 pp.

Geiger, R., 1961. Das Klima der Bodennahen Luftschicht. Vieweg, Braunschweig, 646 pp.

Genderen, J.L. van, 1974. Remote sensing of environmental pollution on Teesside. Environ. Pollut., 6: 220—234.

Georgii, H.W., 1970a. The effects of air pollution on urban climates. In: Urban climate, Techn. Note No. 108, WMO Geneva, pp. 214—237.

Georgii, H.W., 1970b. Das natuerliche Aerosol in reiner und verunreinigter Luft. Schriftenr. Ver. Wasser-, Boden-, Lufthyg., No. 30, p. 13.

Gill, C.J., 1970. The flooding tolerance of woody species. For. Abstr., 31: 671—688.

Glatzel, G., 1974. Analytische Methoden zum Nachweis der Schaedigung von Pflanzen durch Auftausalze. Eur. J. For. Pathol., 4: 52—53.

Glatzel, G. and Krapfenbauer, A., 1975. Streusalzschäden am Baumbestand der Strassen in Wien. Inst. Forstl. Standortforschung Hochschule Bodenkultur Wien, 52 pp.

Goecke, M., 1972. Verpflanzen alter Baeume. Gartenamt, 21: 19—34.

Gollwitzer, G. and Wirsing, W., 1971. Dachflaechen, bewohnt, belebt, bepflantz. Callwey, Muenchen, 136 pp.

Graefe, K. and Schuetze, W., 1966. Staubniederschalgmessungen mit 230 Bergerhoff-Geraeten. Staedtehygiene, Heft 8.

329

Greiner, J. and Gelbrich, H., 1974. Gruenflaechen der Stadt. Grundlagen fuer die Planung, Grundsaetze, Kennwerte, Probleme, Beispiele. Berlin, VEB Verlag, Bauwesen, 192 pp.

Greszta, J. and Olszowski, J., 1972. Einfluss der chemischen Industrieimmissionen auf den Zuwachs von Kiefernbestaenden. Mitt. Forstl. Bundes-Versuchsanst., 97: 163—172.

Grodzinska, K., 1971. Acidification of tree bark as a measure of air pollution in southern Poland. Bull. Acad. Pol. Sci., Ser. Sci. Biol., 19: 189—195.

Gruen, W., 1972. Fassaden, Fenster, Flachdaecher. Deutsche Bauzeitschrift (DBZ), Heft 8: 1467—1487.

Gruen, W., 1975. Muellkompost und Naehrsubstrat fuer Gruen an Fassaden und auf Daechern. Bau-Intern, (1/2): 31—36.

Gutacker, H.W., 1972. Umweltschutz aus der Sicht der Medizin. Korrespondenz Abwasser, 19 (7): 139—145.

Håbjørg, A., 1973a. Air pollution and vegetation. I. Effects of soil and fertilization on growth and development of Alnus incana (L.) Moench and Betula pubescens Ehrh. grown in containers at a SO_2-exposed place in Sarpsborg and at the Agricultural University of Norway in As. Meld. Nor. Landbrukshoegsk., 52, No. 1: 1—23.

Håbjørg, A., 1973b. Air pollution and vegetation. II. Effects of fertilization on growth and development of twenty woody plants grown in industrial areas. Meld. Nor. Landbrukshoegsk., 52, No. 2: 1—14.

Hädrich, F. and Blum, W.E., 1972. Gutachten zur Frage der Salzschaeden an Alleebaeumen insbesondere der Rosskastanie der Stadt Freiburg. Institut fuer Bodenkunde Univ. Freiburg (unpublished).

Hansen, H.H., 1975. Die Pflanze als Sauerstofflieferant. Staatsexamensarbeit (Fachbereich Biologie—Botanik) der Universitaet des Saarlandes, Saarbruecken, 108 pp.

Harley, J.L., 1969. The biology of mycorrhiza. L. Hill, London, 334 pp.

Harney, B.M., McCrea, D.H. and Forney, A.J., 1973. Aerial detection of vegetation damage utilizing a simple 35 mm camera system. J. Air Pollut. Control Assoc., 23/9: 788—790.

Harris, R.W., 1974. Staking landscape trees. Arborist's News, 39: 157—161.

Hasel, K., 1971. Waldwirtschaft und Umwelt. Eine Einführung in die forstwirtschafts-politischen Probleme der Industriegesellschaft. Parey, Hamburg, Berlin, 321 pp.

Hatch, A.B., 1936. The role of mycorrhizae in afforestation. J. For., 34: 22—29.

Hatch, A.B., 1937. The physical basis of mycotrophy in the genus Pinus. Black Rock For. Bull., 6: 168 pp.

Haupt, R., 1973, 1974. Beitrag zum Problem der Laermminderung durch Waldbestaende. Teil 1: Methodische Grundlagen. Teil 2: Die Ergebnisse und ihre Anwendungsmoeglichkeiten. Arch. Naturschutz Landschaftsforsch., 13: 309—327; 14: 61—75.

Hauten, H. van and Stratmann, H., 1967. Experimentelle Untersuchungen ueber die Wirkung von Stickstoffdioxid auf Pflanzen. Schriftenr. Landesanst. Immissions- Bodennutzungsschutz, Essen, Heft 7: 50—70.

Haverbeke, D.F., 1973. Renovating old decisuous windbreaks with conifers. J. Soil Water Conserv., 28: 65—68.

Hayward, H.E. and Long, E.M., 1941. Anatomical and physiological responses of the tomato to varying concentrations of sodium chloride, sodium sulphate and nutrient solutions. Bot. Gaz., 102: 437—462.

Hellpach, W., 1965. Geopsyche. Die Menschenseele unter dem Einfluss von Wetter, Klima, Boden und Landschaft. (7. Aufl.). F. Enke, Stuttgart, 275 pp.

Hennebo, D., 1955. Staubfilterung durch Gruenanlagen. VEB Verlag Technik, Berlin, 79 pp.

Herbst, W., 1965. Filter und Schutzwirkung des Waldes gegen radioaktive und andere Beimengungen der Atmosphaere. Forst- Holzwirt, 20: 216—219.

Herbst, W., 1968. Immissionsschutz durch Baeume und Straeucher. VDI-Nachrichten, 22 (8): 1—2.

Hettche, H.O., 1970. Lungenkrebs und Luftverunreinigung — ein Beitrag zur Epidemologie. Heft 18, Schriftenr. Landesanst. Immissionsbodennutzungsschutz des Landes Nordrhein-Westf., pp. 7—43.

Hicks, P., without year. The care of trees on development sites. Advisory leaflet No. 3. Arboric. Assoc., Northwood, 12 pp.

Hillier, H.G., without year. A tree for every site. A select list of trees and shrubs. Leaflet No. 5. Arboric. Assoc., Worplesdon, 50 pp.

Himelick, E.B., Neely, D. and Crowley, W.R., 1965. Experimental field studies on shade tree fertilization. Ill. Nat. Hist. Survey, Biological Notes 53, 12 pp.

Hoeks, J., 1971. Verbetering van de bodemventilatie bij straatbomen. (5. Rapport of Studiekommissie invloed aardgas op beplantingen), 's Gravenhage, 19 pp.

Hoeks, J., 1972. Effect of leaking natural gas on soil and vegetation in urban areas. PUDOC, Wageningen, 120 pp. (= Agric. Res. Rep., No. 778).

Hoeks, J. and Leegwater, J., 1972. De groei van planten in "gasgrond". (7. Rapport of S.I.A.B.), 's Gravenhage, 16 pp.

Hoffmann, A., 1954. Der Strassenbaum in der Grossstadt unter besondrer Beruecksichtigung der Berliner Verhaeltnisse. Diss. Landwirtsch. Fakultaet der Humboldt-Univ. Berlin, 144 pp.

Holliday, I. and Hill, R., 1969. A field guide to Australian trees, 244 pp.

Hubbard, J.C.E. and Grimes, B.H., 1972. The analysis of coastal vegetation through the medium of aerial photography. Med. Biol. Illus., 22: 182—190.

Huber, B., 1956. Die Saftströme der Pflanzen. Springer, Berlin, Göttingen, Heidelberg, 126 pp.

Huber, B., 1962. Wundheilung bei Pflanzen. Langenbecks Arch. Chir.; 301: 71—79.

Johnson, 1970. The balance of atmospheric oxygen and carbon dioxide. Biol. Conserv., 2: 83—89.

Kadro, A. and Kenneweg, H., 1973. Das Baumsterben auf dem Farb-Infrarotluftbild. Gartenamt, 22: 149—157.

Kayll, A.J., 1963. A technique for studying the fire tolerance of living tree trunks. Publ. Dept. For. Canada, 1012: 1—22.

Keller, M.M., 1968. Der heutige Stand der Forschung ueber den Einfluss des Waldes auf den Wasserhaushalt. Schweiz. Z. Forstwes., 119: 364—379.

Keller, T., 1971a. Die Bedeutung des Waldes fuer den Umweltschutz. Schweiz. Z. Forstwes., 122: 600—613.

Keller, T., 1971b. Auswirkungen der Luftverunreinigungen auf die Vegetation. Ber. Eidg. Anst. Forstl. Versuchwesen, No. 67, Birmensdorf, 11 pp.

Keller, T., 1973a. Die Sauerstoffbilanz der Schweiz. Schweiz. Z. Forstwes., 124: 465—473.

Keller, T., 1973b. Report on the IUFRO meeting "Air pollution effects on forests", Sopron (Hungary), October 9—14, 1972. Eur. J. For. Pathol., 3: 56—60.

Keller, T., 1973c. Zur Phytotoxizitaet staubfoermiger Fluorverbindungen. Staub-Reinhalt. Luft, 33/10: 395—397.

Keller, T., 1974a. The use of peroxidase activity for monitoring and mapping air pollution areas. Eur. J. For. Pathol., 4 (1): 11—19.

Keller, T., 1974b. Die Vegetationsverbleiung durch Bleibenzin. Beitr. Foerderung Biolog. Dynam. Landw. Methode Schweiz, 23 (2): 15—19.

Keller, T., 1974c. Verkehrsbedingte Luftverunreinigungen und Vegetation. Garten Landschaft, 84: 547—550.

Keller, T., 1974d. Ueber die Filterwirkung von Hecken fuer verkehrsbedingte staubfoermige Luftverunreinigungen, insbesondere Bleiverbindungen. Schweiz. Z. Forstwes., 125: 719—735.

Keller, T., 1973. Ueber die schädigende Wirkung des Fluors. Schweiz. Z. Forstwes., 124: 700—706.

Keller, T., 1975a. Zur Phytooxizität von Fluorimmissionen für Holzarten. Mitt. Eidg. Anst. Forstl. Versuchwesen, 51 (2): 303—331.

Keller, T., 1975b. Zur Fluor-Translokation bei Waldbäumen, Mitt. Eidg. Anst. Forstl. Versuchwesen, 51 (3): 335—356.

Kielbaso, J.J., 1975a. Should hardiness zones and location be a part of the ISTC Shade Tree evaluation formula? J. Arboric., 1 (5): 93—97.

Kielbaso, J.J., 1975b. Economic values of trees in the urban locale. Trees Magazine, 34: 9—13; 18.

Kilgore, W.W. and Doutt, R.L. (Editor), 1967. Pest control, biological, physical and selected chemical methods. Academic Press, New York, London, 477 pp.

King, G., 1971. Polyurethane for filling tree cavities. Arborist's News, 36 (2): 13a.

King, G., Beatty, C. and McKenzie, M., 1970. Polyurethane for filling tree cavities. Publication No. 58, University of Massachusetts, Amherst, 6 pp.

Kirchner, H.A., 1967. Grundriss der Phytopathologie und des Pflanzenschutzes. G. Fischer, Jena, 272 pp.

Kirschbaum, U., 1972. Kartierung des natuerlichen Flechtenvorkommens. In: Lufthygienisch. meteorologische Modelluntersuchung in der Region Untermain. RPU, Frankfurt/Main, pp. 76—80.

Kirstensen, K.J., 1959. Temperature and heat balance of soil. Oikos, 10: 103—120.

Klaffke, K., 1972. Funktion und Bedeutung des Gruens in der Stadt. Gartenamt, 21: 261—267.

Knabe, W., 1971. Luftverunreinigungen — Forstlicher Standortfaktor oder abwehrbares Uebel? Hinweise zur Erkennung und Abwehr von Immissionsschaeden im Forstbetrieb. Forstarchiv, 42 (8/9): 172—179.

Knabe, W., 1972. Immissionsbelastung und Immissionsgefaehrdung der Waelder im Ruhrgebiet. Schriftenr. Landesanstalt Immiss. Bodennutzungsschutz Landes Nordrhein-Westf., Heft 26: 83—87.

Knabe, W., 1973a. Luftverunreinigung und Forstpflanzen. VIII. Int. Arbeitstagung forstlicher Rauchschadensachverstaendiger 1972 in Ungarn. Allg. Forstz., 28 (9/10): 196—200.

Knabe, W., 1973b. Zur Ausweisung von Immissionsschutzwaldungen. Forstarchiv. 44 (2): 21—27.

Knabe, W., Brandt, C.S., van Haut, H. and Brandt, J.C., 1973. Nachweis photochemischer Luftverunreinigung durch biologische Indikatoren in der Bundesrepublik Deutschland. Proc. 3rd Int. Clean Air Congr. Duesseldorf, VDI Duesseldorf.

Knapp, R., 1965. The vegetation of North and Central America. Fischer, Stuttgart, 40 + 373 pp.

Knapp, R., 1971. Einfuehrung in die Pflanzensoziologie, 3rd ed. Ulmer, Stuttgart, 388 pp. (with tables on European trees and shrubs and their ecological and growth properties).

Kanpp, R., 1973. The vegetation of Africa. Fischer, Stuttgart, 43 + 626 pp. (including Mediterranean vegetation).

Knapp, R., 1974. (Editor) Vegetation dynamics. Junk, Den Haag, 368 pp.

Knochenhauer, W., 1934. Inwieweit sind die Temperatur — und Feuchtigkeits — messungen unserer Flughaefen repraesentativ? Erf. Ber. Deutsch. Flugwetterd., 9. Folge No. 2.

Koch, W., 1971. Verkehrs- und Schadensersatzwerte von Bäumen, Sträuchern, Hecken und Obstgehölzen nach dem Sachwertverfahren. Pflug und Feder, Bonn, 52 pp.

Kozlowski, T.T., 1971a. Growth and development of trees. Vol. I + II, Academic Press, London, 443 + 514 pp.

Kozlowski, T.T., 1971b. Water needs of trees. Am. Hortic. Mag., 50 (3): 102—106.

Kozlowski, T.T., 1972. Basic principles of tree growth. Arborist's News, 37: 65a—72b.

Kozlowski, T.T. and Keller, T., 1966. Food relations of woody plants. Bot. Rev., 32: 293—382.

Kramer, P.J. and Kozlowski, T.T., 1960. Physiology of trees. McGraw-Hill, New York.

Kratzer, A., 1956. Das Stadtklima, 2. Aufl. Vieweg, Braunschweig, 184 pp. Research translation by U.S. Air Force Cambridge Research Laboratories.

Kräusel, R., 1950. Versunkene Floren. Senckenberg-Buch, No. 25. Kramer, Frankfurt/M, 152 pp.

Krenn, K., 1933. Die Bestrahlungsverhaeltnisse stehender und liegender Staemme. Allg. Forst- Jagdz. (Wien), 51: 50—51; 53—54.

Kreutz, W., 1952. Der Windschutz. Windschutzmethodik, Klima und Bodenertrag. Ardey, Dortmund, 167 pp. + 19 Tafeln.

Kreutz, W. and Walter, W., 1960. Stroemungs- und Erosionsvorgaenge im Bereich von einzelnen und gestaffelten Windschutzanlagen (nach Untersuchungen im Windkanal). Schriftenr. Inst. Naturschutz, Darmstadt, V. 4: 83—103.

Kreutzer, K., 1974. Bodenkundliche Aspekte der Streusalzanwendung. Eur. J. For. Pathol., 4 (1): 39—41.

Krieg, A., 1961. *Bacillus thuringiensis* Berliner. Ueber seine Biologie, Pathogenie und Anwendung in der biologischen Schaedlingsbekaempfung. Mitteil. Biolog. Bundesanstalt, Berlin-Dahlem, 79 pp.

Krieger, M.H., 1973. What's wrong with plastic trees? Science, 179: 446—455.

Kruessmann, G., 1974. Amerikanische Erfahrungen mit Gehoelzen auf ueberschwemmten Baumschulflaechen. Dtsch. Baumsch., 26: 244.

Kühne, H. and Koester, P., 1967. Erdgasschaeden an Strassenbaeumen. Nachrichtenblatt Dtsch. Pflanzenschutzdienstes (Braunschweig), 19: 121—122.

Kurusa, J. and Brause, A., 1974. Gehoelze auf dem Mittelstreifen der Bundesautobahnen. Neue Landschaft, 19: 70—77.

Lagerstedt, H.B., 1971. Tree trunk protection against summer sunscald. Arborist's News, 36: 61—65.

Lamp, W., 1947. Unterwuchungen ueber den Staubgehalt einer vom Krieg teilweise zerstoerten Grosstadt. Diplomarbeit TH Darmstadt.

Lamp, W., 1972. Operation Infrared. RPU Frankfurt/M, 3. Arbeitsbericht, 18—23.

Lampadius, F., 1963. Die lufthygienische Bedeutung des Waldes in ihrer Abhaengigkeit von schaedlichen Raucheinwirkungen auf den Wald. Angew. Meteorol., 4: 248—249.

Lampadius, F., 1968. Die Bedeutung der SO_2-Filterung des Waldes im Blickfeld der forstlichen Rauchschadentherapie. Wissensch. Z. Tech. Univ. Dresden, 17 (2): 503—511.

Landsberg, H.E., 1966. Air pollution and urban climate. Biometeorologie II, Pergamon.

Landsberg, H.E., 1970a. Micrometeorological temperature differentiation through Urbanization. In: Urban Climate, Techn. Note No. 108: 129—136, WMO Geneva.

Landsberg, H.E., 1970b. Man-made climatic changes. Science, 170: 1265.

Lange, O.L. and Schulze, E.D., 1971. Measurement of CO_2 gas-exchange and transpiration in the Beech (*Fagus sylvatica* L.). In: Integrated Experimental Ecology (Ecological Studies 2), pp. 16—28.

Larcher, W., 1973a. Oekologie der Pflanzen. Ulmer, Stuttgart, 320 pp.

Larcher, W., 1973b. Limiting temperatures for life functions in plants. In: H. Precht, J. Christophersen, H. Hensel and W. Larcher (Editors), Temperature and Life. Springer, Berlin, Heidelberg, New York, 195—203.

Lechner-Knecht, W., 1974. Gruene Staedte-gruenes Persian. Baumzeitung, 8: 27—28.

Leh, H.O., 1972. Schaeden an Baeumen durch Auftausalze. Baumzeitung, 6: 22—23.

Leh, H.O., 1973. Untersuchungen ueber die Auswirkungen der Anwendung von Natriumchlorid als Auftaumittel auf die Strassenbaeume in Berlin. Nachrichtenblatt Dtsch. Pflanzenschutzdienstes (Braunschweig), 25 (11): 163—170.

Leh, H.O., 1974. Untersuchungen ueber Toleranzunterschiede von Strassenbaumen gegenueber Auftausal (Natriumchlorid) unter besondrer Beruecksichtigung ihres Aufnahmeverhaltens. Bund Deutscher Baumschulen, Pinneberg, 16 pp.

Leibundgut, H., 1943. Ueber kritische Schneelasten. Schweiz. Z. Forstwes., 94: 61—62.

Leibundgut, H., 1975. Wirkungen des Waldes auf die Umwelt des Menschen. Rentsch., Erlenbach, Zuerich, Stuttgart, 186 pp.

Liddle, M.J., 1975. A selective review of the ecological effects of human trampling on natural ecosystems. Biol. Conserv., 7: 17—36.

Lieth, H., 1964. Versuch einer kartographischen Darstellung der Produktivitaet der Pflanzendecke auf der Erde. Geographisches Taschenbuch, Wiesbaden, pp. 72—80.

Lieth, H., 1972. Ueber die Primaerproduktion der Pflanzendecke auf der Erde. Angew. Bot., 46: 1—37.

Löbner, A., 1935. Horizontale und vertikale Staubverteilung in einer Grosstadt. Veröff. d. Geophys. Inst. d. Universität Leipzig, 7: 53—99.

Lorenz, D., 1972. Investigation of cold-air flows at the Taunus mountains through the use of thermal mapping. RPU Frankfurt/M, 3. Arbeitsbericht, 23—50.

Lorenz, H., 1952. Strasse und Landschaft. Bauverwaltung, 1 (9), 9 pp.

Lorenz, K., 1973. Die acht Todsuenden der zivilisierten Menschheit. Piper, Muenchen, 112 pp.

Lötschert, W. and Koehm, H.J., 1973a. ph-Wert und S-Gehalt der Baumborke in Immissionsgebieten. Oecol. Plant., 8 (3): 199—209.

Lötschert, W. and Koehm, H.J., 1973b. Baumborke als Anzeiger von Luftverschmutzungen. Umsch. Wiss. Tech., 73: 403—404.

Lowry, W.P., 1972. Atmospheric pollution and global climatic change. Ecology, 53 (5): 908—914.

Lundholm, B., 1972. Remote sensing and international affairs. Ambio, 1 (5): 166—173.

Lutz, H.J., 1963. Forest ecosystems, their maintenance, amelioration and deterioration. J. For., 61: 563—569.

Lyr, H., Polster, H. and Fiedler, H.J., 1967. Gehoelzphysiologie. VEB Fischer, Jena, 444 pp.

McCormick, R.A.S.H.L., 1967. Climatic modification by atmospheric aerosols. Science, 56: 1358.

Malcolm, D.C., 1974. Root development and tree stability. In: Tree Growth in the Landscape. Wye College, Kent, pp. 77—82.

Mannhardt, W., 1963. Wald- und Feldkulte. Vol. I and II. Wissenschftl. Buchgesellsch., Darmstadt, 645 + 359 pp.

Martin, A. and Barber, F.R., 1971. Some measurements of loss of atmospheric sulphur dioxide near foliage. Atmos. Environ., 5 (5): 345—352.

Martin, W. and Keller, T., 1974. Ueber den Einfluss verunreinigter Stadtluft auf die Vegetation. Berichte Eidgen. Anst. Forstl. Versuchswesen, No. 134 (= Umwelthygiene, 25 (10): 221—227).

Martini, E., 1952. Keime und Krankheitserreger. In: A. Seybold and H. Woltereck (Editors), Klima- Wetter- Mensch. Quelle und Mayer, Heidelberg, pp. 190—225.

Marx, D.H., 1971. The beneficial relationship between tree roots and mycorrhizae fungi. Arborist's News, 36: 89a—94a.

Materna, J., 1963. Einige Forschungsergebnisse aus der Rauchschadenszone des Erzgebirges. Tagungsbericht III. Bioklimatische Konferenz ueber Fragen der Luftverunreinigungen 1961 in Prag. Naklada telstvi Ceskoslovenske Academie ved Praha, pp. 156—170.

Matthews, W.E., 1974. New advances in tree planting and aftercare. In: Tree Growth in the Landscape. Wye College, Kent, p. 24.

Mayer, R. and Ulrich, B., 1974. Conclusions on the filtering action of forests from ecosystem analysis. Oecol. Plant., 9 (2): 157—168.

Mazek-Fialla, K., 1974. Der Einfluss von Windschutzanlagen auf Schneeansammlung und Schneeschmelze. Allg. Forstz., 29 (49): 1104—1105.

334

Meldau, R., 1956. Handbuch der Staubtechnik I. Duesseldorf.

Mellanby, K., 1972. The biology of pollution. Edward Arnold, London, 60 pp. (= Studies in Biology No. 38).

Melzer, E., 1962. Beitraege zur Wurzelforschung. Diss. Tech. Univ. Dresden, 227 pp.

Metcalf, L.J., 1972. The Cultivation of New Zealand Trees and Shrubs. 292 pp.

Metzger, W., 1976. Gesetze des Sehens. Kramer, Frankfurt a.M., 676 pp.

Meusel, H., Jaeger, E. and Weinert, E., 1965. Vergleichende Chorologie der Zentraleuro-paeischen Flora. Vol. I. + II. Fischer, Jena, 583 + 258 pp.

Meyer, F., 1965. Das Altern der Baeume. Mitt. Dtsch. Dendrol. Ges., No. 62: 59—65.

Meyer, F.H., 1972. Gehoelze in staedtischer Umwelt. Neues Archiv Niedersachsen, 21 (3): 1—24.

Mitscherlich, A., 1969. Die Unwirtlichkeit unserer Staedte. Suhrkamp, Frankfurt/M, 160 pp.

Mitscherlich, A., 1971. Wachstum, Planung, Chaos. In: U. Schulz (Editor), Umwelt aus Beton oder unsere unmenschlichen Staedte: 130—138. Rororo aktuell 1497, Reinbeck.

Mueller-Dombois, D. and Ellenberg, H., 1974. Aims and methods of vegetation ecology. Wiley, New York, 547 pp.

Muench, W.D., 1972. Aenderungen der Artenzusammensetzung von Unkraeutern in Forstkulturen nach Anwendung von Herbiziden. Z. Pflanzenkr. Pflanzenschutz, 79: 485—497.

Munn, R.E., 1970. Airflow in urban areas. In: Urban Climate. Tech. Note No. 108: 15—39, WMO Geneva.

Murphy, R.C. and Meyer, W.E., 1969. The care and feeding of trees. Crown, New York, 165 pp.

Naegeli, W., 1941. Ueber die Bedeutung von Windschutzstreifen zum Schutz landwirt-schaftlicher Kulturen. Schweiz. Z. Forstwes., 11: 265—280.

Naegeli, W., 1943. Untersuchungen ueber die Windverwehungen im Bereich von Wind-schutzstreifen. Mitt. Schweiz. Anst. Forstl. Versuchswes., 23: 223—276.

Naegeli, W., 1954. Die Windbremsung durch einen groesseren Waldkomplex. Ber. 11. Kongr. Int. Verband. Forstl. Forsch. Anst. (Firenze), 240—246.

Naegeli, W., 1965. Ueber die Windverhaeltnisse im Bereich gestaffelter Windschutzstrei-fen. Mitt. Schweiz. Anst. Forstl. Versuchswes., 41: 221—300.

Neely, D., 1971. Healing tree wounds. Arborist's News, 36: 41a—43a.

Neely, D., Himelick, E.B. and Crowley, W.R., 1970. Fertilization of established trees. Ill. Nat. Hist. Survey, 30: 235—266.

Neuwirth, R., 1965. Der Wald als Aerosolfilter. Forst- und Holzwirt, 20.

Neuwirth, R., 1974. Bioklima. In: W. Pehnt (Editor), Die Stadt. Reclam, Stuttgart, pp. 214—237.

Numata, M. (Editor) 1974. The flora and vegetation of Japan. Elsevier, Amsterdam, Kodansha, Tokyo, 294 pp.

Odum, E.P., 1971. Fundamentals of ecology. Saunders, Philadelphia, London, Toronto, 574 pp.

Ortner, W., 1964. Bemessungsgrundlagen für Verkehrssicherheitspflanzungen an Strassen. Str. Tiefbau, 7: 383—391.

Ovington, J.D., 1962. Quantitative ecology and the woodland ecosystem concept. Adv. Ecol. Res. London, 1: 103—192.

Palmer, J.G. and May, C., 1970. Additives, durability and expansion of a urethane foam useful in tree cavities. Plant Dis. Rep., 54 (10): 858—862.

Patoharju, O., 1974. Baumschutzvorschriften der Stadt Helsinki. Baumzeitung, 8: 42; 51.

Pelisek, J., 1974. Changes in the forest stands and soils in Europe. Nature in Focus, No. 19: 3—7.

Peterson, E.K., 1970a. The atmosphere: A clouded horizon. Environment, 12: 32—39.

Peterson, E.K., 1970b. Gas-exchange. Environment, 12: 44—45.

Pfleiderer, H. and Schittenhelm, A., 1952. Klima als Behandlungsmittel. In: A. Seybold and H. Woltereck (Editors), Klima- Wetter- Mensch. Quelle und Mayer, Heidelberg, pp. 170—189.

Piperek, M., 1957. Das Problem der Psychohygiene des Stadtbewohners. In: Gruen und Wasser in der Stadt. Verein deutscher Gewaesserschutz, No. 2: 7—15 (Frankfurt/M).

Pirone, P.P., 1972. Tree maintenance. 4th edn. Oxford University Press, New York, 574 pp.

Pirone, P.P., 1973. Advances in general tree maintenance. Proc. ISTC, 49: 55—60. (Suppl. Arborist's News, 39 (3)).

Pollanschuetz, J., 1968. Erste Ergebnisse über die Verwendung des Infrarot-Farbfilmes in Österreich für die Zwecke der Rauchschadensfeststellung. Centralbl. Gesamte Forstwes., 85: 65—79.

Polster, H., 1950. Die physiologischen Grundlagen der Stofferzeugung im Walde. Untersuchungen ueber Assimilation, Respiration und Transpiration unserer Hauptholzarten. Deutscher Landwirtsch., Muenchen, 96 pp.

Pott, F., 1972. Die Bewertung der Wirkungen von Luftverunreinigungen fuer die Gesundheit des Menschen. BDLA Schr. Reihe No. 13: 41—48, Callway, Muenchen.

Pruen, H., 1975. Bodenphysikalische Einflussnahme auf Substrat- und Bodeneigenschaften durch Schaum- und Wirkstoffe. Neue Landschaft, 20: 128—142.

Pueckler-Muskau, Fuerst von, 1834. Andeutungen ueber Landschaftsgaertnerei. Hallbergsche Verlagshandlung, Stuttgart.

Ranft, H. and Daessler, H.G., 1970. Rauchhaertetest an Gehoelzen im SO_2-Kabinenversuch. Flora, 159: 573—588.

Ranwell, D.W., Winn, J.M. and Allen, S.E., 1973. Road salting effects on soil and vegetation and use of salt for de-icing. Nat. Environ. Res. Counc., London, 24 pp.

Rech, P., 1966. Inbild des Kosmos. 2 Bände. O. Mueller, Salzburg, 610 + 605 pp.

Reichle, D.E., 1970. Analysis of temperate forest ecosystems. In: Ecological Studies Vol. 2, Springer, Heidelberg, New York, Berlin.

Reifferscheid, H., 1950. Staub in Truemmerstaedten. Umsch. Wiss. Tech., 50, Heft 18.

Roberts, B.R., 1971. Trees as air purifiers. Arborist's News, 36: 23a—25a.

Rodin, L.E. and Bazilevich, N.I., 1967. Production and mineral cycling in terrestrial vegetation. English translation by G.E. Fogg. London, Edinburgh.

Roempp, H., 1967. Chemis-Lexikon. 4 Vol. Franckh, Stuttgart.

Röhrig, E., 1967. Wachstum junger Laubholzpflanzen bei unterschiedlichen Lichtverhaeltnissen. Allg. Forst- Jagdz., 138: 239.

RPU (Regionale Planungsgemeinschaft Untermain) 1972. Lufthygienisch-meteorologische Modelluntersuchung in der Region Untermain. 3. Arbeitsbericht, Frankfurt/M, 56 pp.

Rudder, B. de, 1952. Grundzuege der Bioklimatik des Menschen. In: A. Seybold and H. Woltereck (Editors), Klima- Wetter- Mensch. Quelle und Meyer, Heidelberg, pp. 91—169.

Ruge, U., 1971a. Erkennen und Verhindern von Auftausalz-Schaeden an Strassenbaeumen. Nachrichtenbl. Dtsch. Pflanzenschutzdienstes (Braunschweig), 23: 133—137.

Ruge, U., 1971b. Grundlagen fuer Dienstanweisungen zur Ueberwachung von Strassenbaeumen. Gartenamt, 20: 214-218.

Ruge, U., 1972a. Baumsterben durch Auftausalze. Umsch. Wiss. Tech., 72: 60—61.

Ruge, U., 1972b. Bedeutung der Baeume im Stadtbegleitgruen der Grossstaedte. Gartenamt, 21: 267—271.

Ruge, U., 1972c. Ursache des Strassenbaumsterbens und moegliche Gegenmassnahmen. Garten Landschaft, 82: 456—458.

Ruge, U., 1974. Salz- und Gasschaeden an Baeumen und Vegetationsflaechen. Neue Landschaft, 19: 423—425.

Rundel, P.W., 1973. The relationship between basal fire scars and crown damage in giant sequoia. Ecology, 54: 210—213.

Rusch, H.P., 1968. Bodenfruchtbarkeit. Eine Studie biologischen Denkens. Haug, Heidelberg, 243 pp.

Sargent, C.S., 1965. Manual of the trees of North America. 2nd edn. Dover, New York, 934 pp.

Schaller, K., 1975. Synthetische Bodenverbesserungsmittel. Gartenamt, 24: 611—615.

Schauberger, W., 1965. Gruener Bericht. Bad Ischl, 12 pp.

Scheffer, F. and Schachtschabel, P., 1973. Lehrbuch des Agrikulturchemie. 1. Teil: Bodenkunde. Encke, Stuttgart, 448 pp.

Schmid, R., 1972. Sauerstofferzeugung des Waldes. Naturwiss. Rundsch., 2: 74.

Schmidt-Burbach, G.M., 1973. Klimaaenderung im urbanen Bereich. U — Das Tech. Umweltmagazin, 20—24.

Schmucker, T., 1942. Silvae orbis No. 4. The tree species of the northern temperate zone and their distribution. Centre International de Sylviculture Berlin-Wannsee, 156 pp + 250 maps.

Schneider, S., 1974. Luftbild und Luftbildinterpretation. (= Lehrbuch der Allgemeinen Geographie Bd. 11) de Gruyter, Berlin, New York, 530 pp.

Schueepp, W., 1974. Meteorologische Aspekte der Energiebilanz von Ballungszentren. In: Klima und humane Umwelt, CIB-Symposium ueber Bauklimatologie 1974: 61—69. Schweiz. Baudokumentation, Blauen.

Schuerholz, G. and Larsson, J., 1973. Falschfarbenbilder fuer die Vegetationskartierung. Allg. Forst- Jagdz., 144: 111—116.

Schuett, P., 1974. Probleme des Herbizideinsatzes in Waldoekosystemen. Forstwiss. Centralbl., 93 (1): 52—56.

Schulze, E.D., 1970. Der CO_2 Gaswechsel der Buche (Fagus sylvatica L.) in Abhaengigkeit von den Klimafaktoren im Freiland. Flora, 159: 177—232.

Schwerdtfeger, F., 1970. Die Waldkrankheiten. Parey, Hamburg, Berlin, 509 pp.

Sekiguti, T., 1970. Thermal situation of urban areas, horizontally and vertically. In: Urban Climate. Tech. Note No. 108: 137—138, WMO Geneva.

Seybold, A. and Woltereck, H. (Editors), 1952. Klima- Wetter- Mensch. Quelle und Meyer, Heidelberg, 292 pp.

Shigo, A.L. and Wilson, C.L., 1971. Are tree wound dressings beneficial? Arborist's News, 36: 85—88.

Shigo, A.L. and Wilson, C.L., 1972. Discoloration associated with wounds one year after application of wound dressings. Arborist's News, 37: 121—124.

Skye, E., 1968. Lichens and air pollution. A study of cryptogamic epiphytes and environment in the Stockholm region. Acta Phytogeographica Suecica 52, Uppsala, 123 pp.

Slabaugh, P.E., 1974. Renewed cultivation revitalizes sodbound shelterbelts. J. Soil. Water Conserv., 29: No. 2, 4 pp.

Smith, E.M., 1971. Some aspects of the use of anti-desiccants to prevent winter injury. Arborist's News, 36: 37—38.

Smith, S.J., 1972. Relative rate of chloride movement in leaching of surface soils. Soil Sci., 114: 259—263.

Smith, W.H., 1974. Air pollution — effects on the structure and function of the temperate forest ecosystem. Environ. Pollut., 6: 111—129.

Sperber, H., 1974. Mikroklimatisch- oekologische Untersuchungen an Gruenanlagen in Bonn. Diss. Landwirtsch. Fakultät Univ. Bonn, 224 pp.

Stach, W., 1969. Untersuchung ueber die Auswirkung der Winterstreuung und anderer ernaehrungsphysiologischer Faktoren auf die Strassenbaeume der Hamburger Innenstadt. Diss. Univ. Hamburg, Mathe. Naturwiss. Fakultaet, 100 pp.

Stearns, E.W., 1949. Ninety years change in a northern hardwood forest in Wisconsin. Ecology, 30: 350—358.

Steiner, H., 1957. Arthropoden des Apfelbaumes. Ihre jahreszeitliche Verteilung und Moeglichkeiten zur Ermittlung ihres Schaedlichkeits- und Nuetzlichkeitsgrades. 14.

Verhandlungsbericht Dtsch. Ges. Angew. Entomologie, Parey, Hamburg, Berlin.

Steinhuebel, G., 1961/62. Rauchhaerte der Immergruenen. Mitt. Dtsch. Dendrol. Ges., 62: 71—76.

Stellingwerf, A., 1973. De technische aspecten van de Infraroodlicht-Fototechniek. S.I.A.B. 6. Rapport, 7—9.

Steubing, L. and Klee, R., 1970. Vergleichende Untersuchungen zur Staubfilterwirkung von Laub- und Nadelgehoelzen. Angew. Bot., 44: 73—85.

Steubing, L. and Klee, R., 1972. Flechtenexplantate. In: Lufthygienische Modelluntersuchung in der Region Untermain, 4. Arbeitsbericht, RPU Frankfurt/M, 81 pp.

Stoefen, D., 1974. Blei als Umweltgift: Die verdeckte Bleivergiftung, ein Massenphaenomen. Schroeder, Eschwege, 214 pp.

Straehler, A.N. and Straehler, A.H., 1973. Environmental Geoscience, Interactions between natural systems and man. Hamilton, Santa Barbara, California.

Strassburger, E., 1971. Lehrbuch der Botanik fuer Hochschulen. (Neubearbeitung von D. von Denffer). Fischer, Stuttgart, 872 pp.

Studiekommissie Invloed Aardgas op Beplantingen, S.I.A.B., 1. Rapport April, 1968, 22 pp.; 2. Rapport September 1968, 6 pp.; 3. Rapport Maerz 1969: Methoden ter verbetering van de samenstelling van de bodenlucht bij straatbomen, 30 pp.; 4. Rapport, Maerz 1970: Voorkoming en bestrijding van aardgasschade aan beplantingen, 16 pp.

Tattar, T.T., 1975. Presymptomatic detection of shade tree diseases by the use of nondestructive bioelectrical techniques. Proc. Tree Wardens, Arborists and Utilities Conf. (1975), Univ. Massachusetts, Dept. Agric. and County Extension Service, 43—45.

Taylor, G.R., 1970. The Doomsday Book. Thames and Hudson, London. 1971, Deutsch: Das Selbstmordprogramm, Fischer Tachenbuch No. 1369, Frankfurt/M, 329 pp.

Thedic, J., 1959. Une enquête sur la securité routière et les plantations d'alignement. Revue générale des routes et des aérodromes, No. 335.

Thiele, A., 1974. Luftverunreinigung und Stadtklima im Grossraum Muenchen, insbesondere in ihrer Auswirkung auf epixyle Testflechten. Ein Beitrag zum Problem des Umweltschutzes. F. Duemmler, Bonn, 175 pp.

Thomas, G.W. and Swoboda, A.R., 1970. Anion exclusion effects on chloride movement in soil. Soil Sci., 110: 163—166.

Trautmann, W., 1966. Erlaeuterungen zur Karte der potentiellen natuerlichen Vegetation der Bundesrepublik Deutschland 1 : 200,000. Blatt 85 Minden. Schriftenr. Veget. kunde, Bundesanst. Vegetationskunde, Naturschutz Landsch. pflege, Heft 1, Bad Godesberg, 137 pp.

Trene, M., 1954, 1955. Ueber Kondensationsvorgaenge im Boden. I. und II. Sitzungsber. Dtsch. Akad. Landbauwiss. Berlin, 3 (6) und 4 (14).

Treshow, M. and Stewart, D., 1973. Ozone sensitivity of plants in natural communities. Biol. Conserv., 5 (3): 209—214.

Trimble, G.R., Reinhart, K.G. and Webster, H.H., 1963. Cutting the forest to increase water yields. J. For., 61: 635—640.

Troll, W., 1973. Allgemeine Botanik. Enke, Stuttgart, 994 pp.

Tschermak, L., 1950. Waldbau auf pflanzengeographischer und oekologischer Grundlage. Springer, Wien, 722 pp.

Tukey, H.B., 1952. The uptake of nutrients by leaves and branches of fruit trees. Rep. XIIIth Int. Hortic. Congr., London, pp. 297—306.

Turner, H., 1968. Der heutige Stand der Forschung ueber den Einfluss des Waldes auf das Klima. Schweiz. Z. Forstwes., 119: 335—352.

Tüxen, R., 1937. Die Pflanzengesellschaften Nordwestdeutschlands. Mitt. Florist. Soziolog. Arb. Gemeinsch. Niedersachsen, Heft 3, 170 pp.

Tüxen, R., 1956. Die heutige potentielle natuerliche Vegetation als Gegenstand der Vegetationskartierung. Angew. Pflanz. Soziologie, 13: 4—52, Stolzenau/Weser (Ber. Dtsch. Landeskunde, 19: 200—246, Remagen, 1958).

Tüxen, R. (Editor), 1961. Pflanzen und Pflanzengesellschaften als lebendiger Bau- und Gestaltungsstoff in der Landschaft. Angew. Pflanz. Soziologie, No. 17, Stolzenau/ Weser, 177 pp.

U.S. Dept. Int., Tree Preservation Bull. No. 1, Washington.

U.S. Dept. Int. 1963. Tree Preservation Bull. No. 3, Washington.

Valen, L. van, 1970. Gas-exchange. Environment, 12: 39—40.

Vester, G., 1972. A metabolic study of flooding tolerance in trees. Doctoral Thesis, University of Munich, 200 pp.

Vins, B. and Mrkva, R., 1972. Zuwachsuntersuchungen in Kiefernbestaenden in der Umgebung einer Duengerfabrik. Mitt. Forstl. Bundes-Versuchsanst., 97: 173—193.

Voelger, K., 1972. Principle and method of infrared measuring flights. In: RPU 3. Arbeitsbericht, 13—18.

Vogellehner, D., 1972. Gehoelze im Licht der pflanzlichen Evolution. Mitt. Dtsch. Dendrol. Ges., 65: 65—75, Schaper, Hannover.

Vogt, H. and Amelung, W. (Editors), 1952. Einfuehrung in die Balneologie und Klimatologie. Springer, Berlin, 266 pp.

Wagner, F., 1968. Cabling and bracing. Arborist's News, 33: 1—5.

Wagner, F., 1969. Cabling and bracing techniques. VLth Int. Shade Tree Conf., August 10—14, 1969, Proc. Annu. Meet., Ohio, Worster, pp. 203—211.

Walter, B., Bastgen, D., Pantenburg, G. and Koch, W., 1974. Einwirkungen von Gas und Auftausalzen auf Boden und Pflanze. Gartenamt, 23: 578—581.

Walter, H., 1950. Einfuehrung in die Phytologie. I. Grundlagen des Pflanzenlebens. Ulmer, Stuttgart, 491 pp.

Walter, H., 1960. Einfuehrung in die Phytologie. III. Grundlagen der Pflanzenverbreitung. 1. Teil: Standortlehre (Analytisch-oekologische Geobotanik). Ulmer, Stuttgart, 566 pp.

Walter, H., 1968. Die Vegetation der Erde in oeko-physiologischer Betrachtung. Band 2. Die gemaessigten und arktischen Zonen. Fischer, Jena, 1001 pp.

Walter, H., 1970. Vegetationszonen und Klima. Ulmer, Stuttgart.

Walter, H., 1973. Die Vegetation der Erde in oeko-physiologischer Betrachtung. Band 1. Die tropischen und subtropischen Zonen (3. Aufl.). Fischer, Jena, 743 pp. Vegetation of the earth in relation to climate and the eco-physiological conditions. Translated from the second German edition by J. Wieser. The English University Press, London. Springer, New York, Heidelberg, Berlin, 237 pp.

Walter, H. and Lieth, H., 1960—1967. Klimadiagramm-Weltatlas. Fischer, Jena, 9000 Diagramme, 33 Hauptkarten, 22 Nebenkarten.

Walter, H. and Straka, H., 1970. Einfuehrung in die Phytologie III/2 Arealkunde, floristisch-historische Geobotanik. Ulmer, Stuttgart, 478 pp.

Walter, H., Harnickell, E. and Mueller-Dombois, D., 1975. Klimadiagramm-Karten der einzelnen Kontinente und die oekologische Klimagliederung der Erde. Eine Ergaenzung zu den Vegetationsmonographien. Fischer, Stuttgart, 36 pp.

Weck, J., 1956. Die Waelder der Erde. Springer, Berlin, Goettingen, Heidelberg, 152 pp.

Weck, J., 1959. Regenwaelder, eine vergleichende Studie forstlichen Produktionspotentials. Z. Erdkunde, 90: 11—37.

Weickmann, L. and Ungeheuer, H., 1952. Grundlagen der Klima- und Wetterkunde. In: A. Seybold and H. Woltereck (Editors), Klima - Wetter - Mensch. Quelle und Meyer, Heidelberg, pp. 11—90.

Weidensaul, T.C., 1973. Are trees efficient air purifiers? Arborist's News, 38: 85—89.

Welte, D.H., 1970. Organischer Kohlenstoff und die Entwicklung der Photosynthese auf der Erde. Naturwissenschaften, 57 (1): 17—23.

Wendorff, G.B.V., 1974. Schuetzt der Wald vor fluglaerm? Gedanken zum forstlichen Laermschutz. Forstarchiv, 45 (1): 6—11.

Wentzel, K.F., 1960. Die Wirkungen des Waldes auf Mensch und Umwelt. Landwirtsch.-Angew. Wiss., No. 107: 140—168.

Wentzel, K.F., 1963. Waldbauliche Massnahmen gegen Immissionen. Allg. Forstz., 18: 101—106.

Wentzel, K.F., 1965. Fluorhaltige Immissionen in der Umgebung von Ziegeleien. Staub, 25: 121—124.

Wentzel, K.F., 1969. Empfindlichkeit und Resistenzunterschiede der Pflanzen gegenueber Luftverunreinigungen. Air Pollution. Proc. 1st European Congr. Influence of Air Pollution on Plants and Animals: PUDOC Wageningen, pp. 357—370.

Wentzel, K.F., 1971. Habitus-Aenderung der Waldbaeume durch Luftverunreinigung. Forstarchiv, 41: 165—172.

Wert, S.L., 1972. Detecting and evaluating air pollution damage to forest stands in the United States by using large scale color aerial photography. Mitt. Forstl. Bundes-Versuchsanst., No. 97: 281—293.

Whittaker, R.H., 1970. Communities and Ecosystems. London.

Whittaker, E. (Editor), 1973. Ordination and classification of plant communities. Junk, Den Haag, 738 pp.

Wilks, J.H., 1969. Tree surgery: Bracing. Arboric. Assoc. J., Vol. (8): 201—206.

Wilmanns, O., 1973. Oekologischen Pflanzensoziologie. Quelle und Meyer, Heidelberg, 288 pp. (= Uni-Taschenbuecher, 269).

Wood, F.A. and Coppolino, J.B., 1972. The influence of ozone on deciduous forest tree species. Mitt. Forstl. Bundes-Versuchsanst., No. 97: 233—254.

Zak, B., 1964. Role of mycorrhiza in root disease. Annu. Rev. Phytopathol., 2: 377—392.

Zimmermann, M.H. and Brown, C.L., 1971. Trees, structure and function. Springer, New York, Heidelberg, Berlin, 336 pp.

Zweiacker, E., 1975. Stroemungsvorgaenge in der erdnahen Atmosphaere. Dtsch. Archtekt. Blatt, 7: 27—28.

Zweimann, B., Slavin, R.G., Feinberg, R.J., et al., 1972. Effects of air pollution on asthma: A review. J. Allergy Clin. Immunol., 50 (5): 305—314.

INDEX

348

bertiana 39, 321; *montana mughus*
108; *monticola* 39, 321; *mugo* 73,
138; *nigra* 29, 39, 106, 107, 321;
palustris 29, 39, 61; *pinea* 47; *ponderosa* 29, 39, 321; *radiata* 70; *resinosa* 39, 321; *rigida* 321; *strobus* 29,
56, 66, 68, 108, 321; *sylvestris* 27, 29,
30, 38, 39, 41, 55—58, 73, 79, 106—
108, 161, 321; *taeda* 27, 39, 61, 65
Pioneer woods 73, 100
Pipes in the root zone *234*
Pith 17, *18, 19*, 30
Plane tree 20, 95, 228, 242
Plant association 70; class 70; dynamic-
genetic classification 70; formation 70;
sociology 70, 72; subassociation 70;
substitute community 71; succession
71; union 70; variant 70
Planting of trees 243, *244—351*; drill 93;
hole 243, *244;* staking *247—249;*
street trees 244, *245—247*, 250; time
243; transplanting large trees 253, 254;
tree mover 255, *256, 257*; watering
249, *250*
Plasma 20
Plastic trees 169
Platanus 65; *acerifolia* 106, 114; *hybrida*
29, 99, 319; *occidentalis* 29, 41, 319;
orientalis 99, 319
Plum 106, 116
Podocarpaceae 13
Podocarpus 65, 182
Polymeric linear colloids 271
Polysaccharides 270
Polyurethane 211
Pomegranate 4
Poplar 14, 19, 27, 68, 72, 95, 106
Populus 29, 38, 53, 61, 65; *alba* 38, 41,
115, 319; *angustifolia* 56; *balsamifera*
320; *berolinensis* 154; *canadensis* 154;
canescens 114, 320; *deltoides* 251,
320; *euramericana* cv. Sacrau 27; *nigra*
38, 41, 56, 73, 115, 320; *nigra italica*
115; *sargentii* 320; *serotina* 251; *tre-
mula* 38, 41, 71, 73, 100, 114, 320;
tremuloides 39, 115, 320; *trichocarpa*
320
Potential natural vegetation 72
Precipitation 77
Preservation of trees 175—242
Pretoria 77
Preventive bud 31
Prior-linden *211*

Protecting trees against damage from construction work 217—242; aerating
system *225, 268;* already filled over
227; filling up the root areas 222; overfilled trees *224;* technique of applying
fills *223;* trees in concrete 228
Protecting trees on building sites 238,
239
Protection effects of forests 79
Protective grating *214, 246*
Protective pipe 193, *194*
Protective regulations for trees 120
Protoplasm 17, 23
Pruning of trees 180—190; how to prune
183, 184; obsolete pruning 24, 51,
181, 182; peeling bark 188; thinning
pruning *183;* when to prune 180;
wound dressing 186; wound healing
184; wound treatment 184—186
Prunus avium 29, 41, 71, 114; *padus* 61,
71, 114, 115; *persica* 62; *serotina* 62,
114, 115; *spinosa* 71, 114, 115; *sub-
hirtella* 62; *virginiana* 251
Pseudolarix 14
Pseudotsuga 65; *douglasii* 161; *menziesii*
29, 56, 115, 321; *taxifolia* 27, 39, 41,
321
Psilophytes 10
Psychic resonance 170
Psycho-physical condition of the city
dweller 169
Pterocarya fraxinifolia 154
Pyracantha coccinea 114
Pyrus communis 29

Querceto—Carpinetum medioeuropaeum
71; *typicum* 71
Quercus 38, 53, 65, 115; *alba* 29, 39,
320; *bicolor* 320; *borealis* 27, 29, 108,
115; *borealis maxima* 39, 41; *calliprinos* 6; *coccinea* 320; *ilex* 47; *lyrata*
320; *macrocarpa* 29, 99, 251, 320;
nigra 320; *palustris* 61, 251, 320;
pedunculata 320; *petraea* 29, 61, 106,
320; *phellos* 61; *pseudoturneri* 29;
robur 29, 56, 58, 61, 71, 99, 107, 108,
114, 115, 154, 161, 320; *rubra* 62, 99,
114, 115, 251, 320; *sessiliflora* 71,
115; *shumardii* 61; *stellata* 41; *suber*
41; *velutina* 320

Radiation balance in Aachen 38, *148,
149*

INDEX OF TREES — LATIN NAMES

354

Ginkgo biloba 13, 14, 29, 65, 99, 108, 251
Gleditsia 64; triacanthos 29, 61, 251, 319
Glyptostrobus 14
Gymnocladus 64

Halimodendron 114
Hamamelis mollis 251
Hedera helix 319
Hippophae rhamnoides 64, 73, 114, 115

Ilex aquifolium 61, 154; decidua 61; opaca 319

Juglans 53; cinerea 319; nigra 29, 62, 251, 319; regia 107, 319
Juniperus 38; communis 29, 73, 321; horizontalis 41; sabine 106; virginiana 29, 115, 321

Keteleeria 14

Laburnum 64
Larix 65, 115; decidua 29, 38, 39, 41, 56, 61, 73, 106—108, 161, 321; kaempferi 29; laricina 29, 39, 61, 321; leptolepis 27, 108, 321
Lauraceae 14
Laurus 6
Lebachiaceae 13
Leguminosae 65
Lepidodendron 12, 14
Lepidosigillaria whitei 10
Libocedrus chilensis 61; decurrens 321
Ligustrum vulgare 106, 115
Liquidambar styraciflua 29, 61, 319
Liriodendron 65; tulipifera 29, 39, 41, 319
Lonicera xylosteum 114; halimifolium 114; ledebouri 154; maackii 154
Lycopodium 12, 14

Maclura pomifera 29
Magnolia acuminata 319; grandiflora 319; soulangeana 62
Malus 251; baccata 115; "Dolgo" 62; hybrid 27; silvestris 29; spec. 62, 115
Metasequoia 65
Moraceae 14
Morus alba 41, 62; spec. 115
Musaceae 15
Myrceugenella apiculata 61

Myrceugenia exsucca 61
Myrica gale 64

Nerium oleander 47
Nothofagus antarctica 61; dombeyi 61; pumilo 61
Nyssa aquatica 61; sylvatica 29, 243, 319

Olea europaea 47, 319
Oxidendron 243

Pachysandra 241
Parthenocissus 126
Paulownia tomentosa 41
Phellodendron amurense 41
Philadelphus pubescens 154
Picea 65; abies 12, 27, 29, 39, 41, 47, 55, 56, 58, 62, 106—108, 114, 138, 161, 321; engelmannii 321; glauca 39, 115, 321; homolepis 108; mariana 321; omorica 29; pungens 62, 115, 321; rubens 39; sitschensis 39, 321
Pinaceae 13
Pinus 61, 65; aristata 321; austriaca 108; banksiana 29, 39, 41, 66, 321; cembra 39, 47, 56, 321; contorta 61, 321; densiflora 48; enchinata 27, 321; flexibilis 321; insignis 77; jeffreyi 321; lambertiana 39, 321; montana mughus 108; monticola 39, 321; mugo 73, 138; nigra 29, 39, 106, 107, 321; palustris 29, 39, 61; pinea 47; ponderosa 29, 39, 321; radiata 70; resinosa 39, 321; rigida 321; strobus 29, 56, 66, 68, 108, 321; sylvestris 27, 29, 30, 38, 39, 41, 55—58, 73, 79, 106—108, 161, 321; taeda 27, 39, 61, 65
Platanus 65; acerifolia 106, 114; hybrida 29, 99, 319; occidentalis 29, 41, 319; orientalis 99, 319
Podocarpaceae 13
Podocarpus 65, 182
Populus 29, 38, 53, 61, 65; alba 38, 41, 115, 319; angustifolia 56; balsamifera 320; berolinensis 154; canadensis hybr. 154; canescens 114, 320; deltoides 251, 320; euramericana cv. "Sacrau" 27; nigra 38, 41, 56, 73, 115, 320; nigra italica 115; sargentii 320; serotina 251; tremula 38, 41, 71, 73, 100, 114, 320; tremuloides 39, 115, 320; trichocarpa 320;

Please see also the tree list "Insect and disease control guide for trees and shrubs" 282—292

INDEX OF TREES — ENGLISH NAMES

Pomegranate 4
Poplar 14, 19, 27, 68, 72, 95, 106
Psilophytes 10

Red bud 109
Red oak 27, 72, 107
Resinous woodpine 27

Sassafras 243
Scarlett oak 109
Scotch pine 27, 56
Seedferns 14
Shingle oak 109
Shortleaf pine 27
Silver birch 56
Sorrel tree 243
Sour-wood 243
Spruce 45, 56, 57, 63, 66, 68, 72, 74—76,
    78, 104, 106, 107, 161, 182
Sugar maple 27, 109
Sweet gum 109, 243
Swiss pine 56, 66
Sycamore 61, 66, 109
Sycamore maple 95

Tamarisk 116
Tree of heaven 291
Tulip poplar 109
Tulip tree 243
Tupelo 243

Vine 5

Walnut 20, 182, 228, 243
White ash 109
White birch 108
White fir 66
White fringe tree 243
White oak 109
White pine 72, 73, 106, 107
White poplar 56
Willow 19, 107

Yellow wood 182, 243
Yew 106, 107

Please see also the tree list "Insect and
disease control guide for trees and shrubs"
282—292